THE ENVIRONMENT IN THE NEWS

SCIENCE NEWS FLASH

The Environment in the News
Genetics in the News
Medicine in the News
Water in the News

THE ENVIRONMENT IN THE NEWS

YAEL CALHOUN

To Alex, Ben Isaac, and Sam—three more habitat savers.

The Environment in the News

Chelsea House
An imprint of Infobase Publishing
132 West 31st Street
New York NY 10001

ISBN-13: 978-0-7910-9253-8
ISBN-10: 0-7910-9253-4

Library of Congress Cataloging-in-Publication Data
Calhoun, Yael.
The Environment in the news / Yael Calhoun.
p. cm. — (Science news flash)
Includes bibliographical references and index.
ISBN 0-7910-9253-4 (hardcover)
1. Environmental responsibility. 2. Environmental management. 3. Endangered species. 4. Air pollution. 5. Climatic changes. I. Title. II. Series.
GE195.7.C35 2007
363.7—dc22 2006035158

Chelsea House books are available at special discounts when purchased in bulk quantities for businesses, associations, institutions, or sales promotions. Please call our Special Sales Department in New York at (212) 967-8800 or (800) 322-8755.

You can find Chelsea House on the World Wide Web at http://www.chelseahouse.com

Text design by Annie O'Donnell
Cover design by Ben Peterson

Printed in the United States of America

Bang EJB 10 9 8 7 6 5 4 3 2 1

This book is printed on acid-free paper.

Contents

Endangered Species

WHAT'S IN THE NEWS?

In April of 2005, the headlines were abuzz with the news: after 60 years of believing that the ivory-billed woodpecker was extinct, birders were thrilled to discover that the bird is still alive (Figure 1.1). Birders were not the only ones who rejoiced. Biologists and conservationists who had been working for more than 50 years to protect the birds' last remaining habitat were ecstatic. The race to save the ivory-billed edwoodpecker stands at the beginning of what has grown to be a worldwide struggle to save **endangered species**.[1]

In May of 2005, a headline announced that a new species of monkey—the highland mangabey—was discovered in Tanzania. It was immediately listed as an endangered species.

Another headline item is the controversy surrounding the revision of the 1973 Endangered Species Act. It was last updated in 1988, and since then Congress has not been able to agree on what changes should be made. Some argue that the act prevents economic growth because it is too restrictive. Others believe that science indicates stronger measures are needed to protect endangered species and the habitats in which they live. The future of the Endangered Species Act, arguably one of our strongest tools for protecting species, will continue to be in the news as the scientists and policy makers debate its future.

How does all this talk of endangered species affect you? We like good news, and the fact that the bald eagle and the gray whale are making a comeback is good news. However, unless you are heading to Africa or Arkansas, does it matter that a new species of monkey has been discovered or that a bird thought to be long extinct has been sighted? Should you care that wolf populations are recovering after being almost being totally eradicated in the lower 48 states? Or that Japan, Iceland, and Norway still hunt whales, some of which are endangered, in total disregard to the international **moratorium** on whaling?

Figure 1.1 By the end of the twentieth century, the ivory-billed woodpecker was thought to be extinct due to heavy logging and hunting. Then, in 2005, a researcher at Cornell University claimed that the bird was sighted in eastern Arkansas. The ivory-billed woodpecker *(illustrated above)* has a shiny blue/black exterior with white markings on its neck and wings.

Many scientists and policy makers all over the world believe it does matter to us all—and for many compelling reasons that will be discussed.

THE SCIENCE

What Is an Endangered Species?

A plant or animal is considered threatened or endangered when its numbers get so low that the species is in danger of becoming extinct. Extinct means, quite simply, that no more of that species exists anywhere on the planet.

An endangered-species designation involves an objective process. In the United States, a plant or animal species is designated by the U.S. Department of Fish and Wildlife as threatened or endangered when a review process determines that the plant or animal is in danger of becoming extinct. The plant or animal, and its habitat, is then afforded a specific set of protections under the current U.S. Endangered Species Act.[2] Recent measures have been introduced to seriously weaken the protective nature of the Act.

On a global level, endangered species are designated by the World Conservation Union (IUCN). The IUCN maintains a list of all the plants and animals it considers endangered. The list is called the Red List of Threatened Species.

Do All Species Become Extinct?

People who argue that extinction is a natural process are correct. Five major extinctions reaching back millions of years have occurred in our planet's history—the last one was 65 million years ago and resulted in the end of all dinosaurs. In fact, about 99% of all the species that have ever lived on the planet are now extinct.

The problem with current extinctions is twofold: First is the rate at which extinctions are happening, and second is the time needed for the planet to recover from mass extinctions. Some scientists estimate the rate may have accelerated by 100 to as much as 1,000 times the natural background rate.[3,4] As Harvard biologist and Pulitzer Prize winner E.O. Wilson explained, it has required tens of millions of years to recover from each of these mass extinctions. Wilson stated, "These figures should give pause to anyone who believes that what *Homo sapiens* destroy, Nature will redeem. Maybe so, but not within any length of time that has meaning for contemporary humanity."[5]

How Do Endangered Species Fit Into Our World?

The key to understanding why endangered species are in the news lies in discovering how living things fit together into a bigger picture. The big picture, in this case, includes not only the particular type of plant or animal but also the special part of the world in which it lives. Removing parts of the picture can trigger ripple effects that we may observe or some we do not yet understand. This way of thinking is relatively new to science. Since the 1800s, people from Henry David Thoreau to Teddy Roosevelt to John Muir have been speaking about conserving nature. However, not until the 1960s did science begin to provide substantial proof that living and nonliving pieces of our planet were connected, and that human activity was having a negative impact.

What Is an Ecosystem?

Plants and animals need certain things to live. They need a place to find food, water, and oxygen, and a safe place to reproduce. In fact, many plants and animals depend on each other for survival. For example, all animals breathe oxygen created by plants, and many also eat the plants. Some plants rely on animals for **pollination** or spreading their seeds. Some animals eat other animals. All these pieces—the living things and the nonliving elements, such as water and soil and air—create a system that allows life to continue. This self-sustaining group of organisms and the physical characteristics (soil, minerals, water, and weather) that make up their surroundings are an **ecosystem**.[6] "Ecosystem" has the same Greek root as "ecology"— *oikos*, meaning "house" or "home."

The major ecosystems on our planet include ocean, freshwater, and terrestrial (land) ecosystems. Each of these major ecosystems can be divided into smaller ecosystems; for example, a rain forest in Brazil, a coral reef in Australia, an alpine meadow in Montana, a salt marsh in New England, or a desert in Arizona. Even a **terrarium** made from an inverted soda bottle is a miniecosystem if it is self-sustaining, meaning that you do not need to add anything to it. Well, one thing must be added—Would the species living in the terrarium survive in a dark closet? An ecosystem needs energy, which it gets from the Sun, to connect the parts.

What Connects the Different Parts of an Ecosystem?

Inside a soda-bottle terrarium may be some soil, a plant, an earthworm, and water. The plant uses the soil for support and nutrients, while also using the water for **photosynthesis**. Dead leaves **decompose** into organic matter. The worm eats the organic matter and excretes casts, which provide available nutrients to the plant. This simple example shows that organisms connect to their environment through the flow of energy and nutrient cycles.[7]

Energy Flow

How does energy enter an ecosystem? Our entire planet runs on solar power. Plants are autotrophs, or primary producers, because they capture energy from the Sun through photosynthesis. Animals, or heterotrophs, cannot capture energy directly from the Sun, so they must eat plants to obtain energy. Plant eaters, such as deer and rabbits, are called primary consumers. Animals such as coyotes and spiders, which eat other animals, are called secondary consumers. Top predators, such as sharks, wolves, and hawks, which eat other predators, are known as tertiary consumers. This chain of dependency for food is called a **food chain** (Figure 1.2).

Each level in the food chain is called a trophic level. News stories often highlight what happens when poisons or toxins at the bottom of the food chain accumulate in animals at the top. Consider that a family of barn owls can eat 1,000 rodents in a nesting

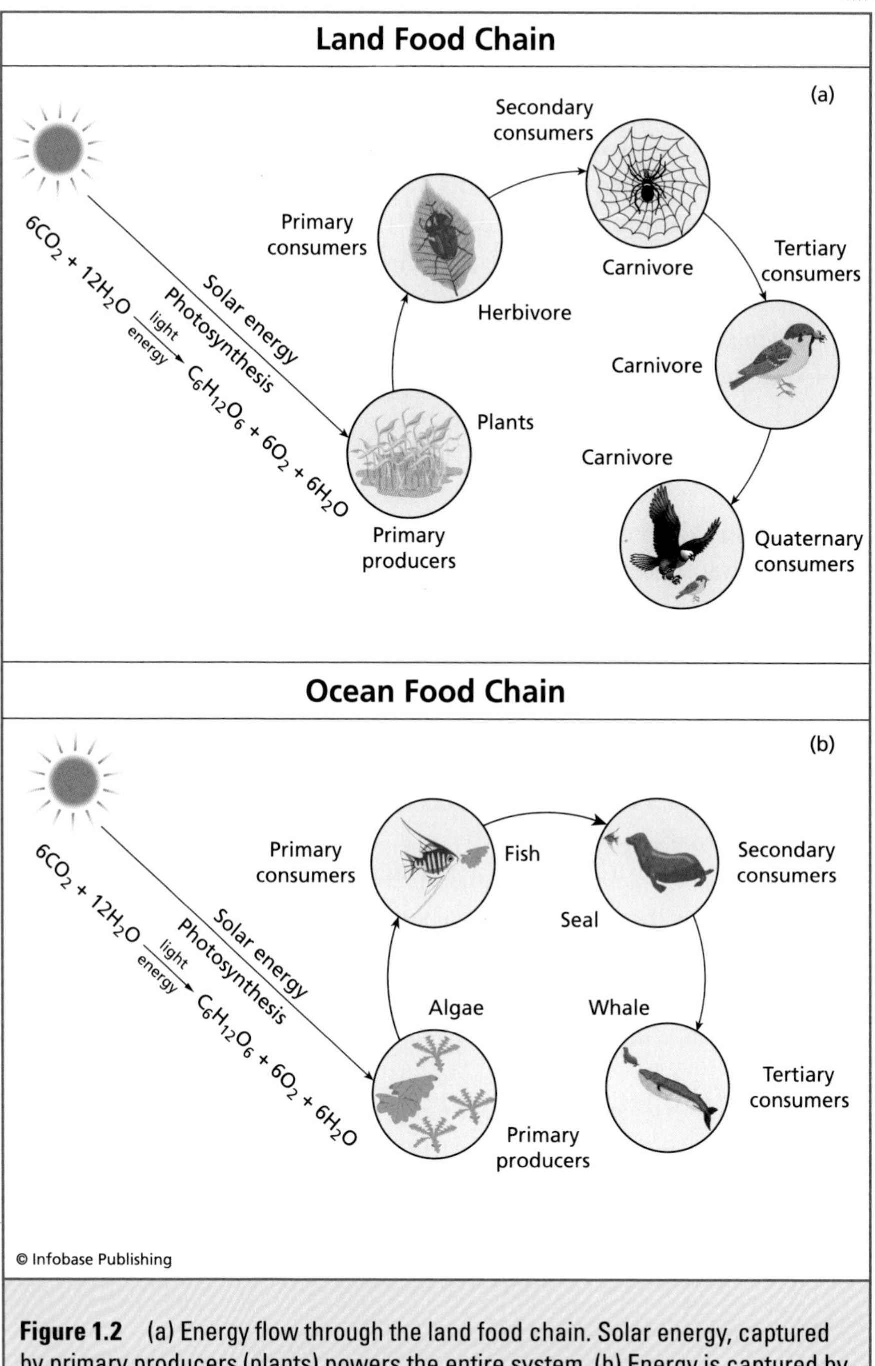

Figure 1.2 (a) Energy flow through the land food chain. Solar energy, captured by primary producers (plants) powers the entire system. (b) Energy is captured by algae (phytoplankton) to form the base of the ocean food chain.

season. If a farmer poisons the rodents and the owls eat them, the toxins can accumulate in the owls. Similarly, mercury can **bioaccumulate** in large predator fish, such as tuna, which eat many smaller fish contaminated with the heavy metal. When humans eat tuna, they can be eating toxic mercury, too.

Humans, with the exception of vegetarians, are at the top of the food chain. If you had a hamburger with cheese for lunch, how many food chains have you eaten from today? Because many animals eat from more than one food chain, scientists call these energy connections food webs.

Nutrient Cycles

What would Earth be like if the dead plants and animals did not decompose? For one thing, people would be walking on deep piles of dead organic material. For another, new plants and animals would have no minerals or nutrients available for growth, because they would be locked up in the dead material. The reason we do not have to deal with such piles of material is that our planet has billions of organisms that break down the organic matter.

Each ecosystem supports its own species of decomposers (bacteria, fungi, or earthworms) that release minerals and nutrients back into the soil and water (Figure 1.3). These materials then are used by new plants and animals as they use energy to metabolize to grow and repair cells. Think about it—the calcium in your milk this morning may well have been part of a *Tyrannasaurus rex* bone from 70 million years ago.

How Do Species Affect Each Other?

Another connection can impact an ecosystem—the way in which species interact with each other.[8] Scientists have learned through observation and study that when one species is disturbed, the rest of the ecosystem can be affected. "Second-order consequences of species loss" refers not only to the disappearing species but also to the effect the loss has on the rest of the ecosystem.[9]

The term "keystone species" expresses the concept of other species being dependent on one particular species. An organism

is considered a keystone species when its removal drastically changes the ecosystem. Scientists have had several opportunities to observe this change. One example was when sea otter populations were greatly reduced, the sea urchin population on which they fed greatly increased. The exploding sea urchin population then overgrazed the kelp (giant seaweed) forests, thus displacing all the fish species that had lived there.[10] The way in which species connect underscores John Muir's observation that "When we try to pick out anything by itself, we find it hitched to everything else in the Universe."[11]

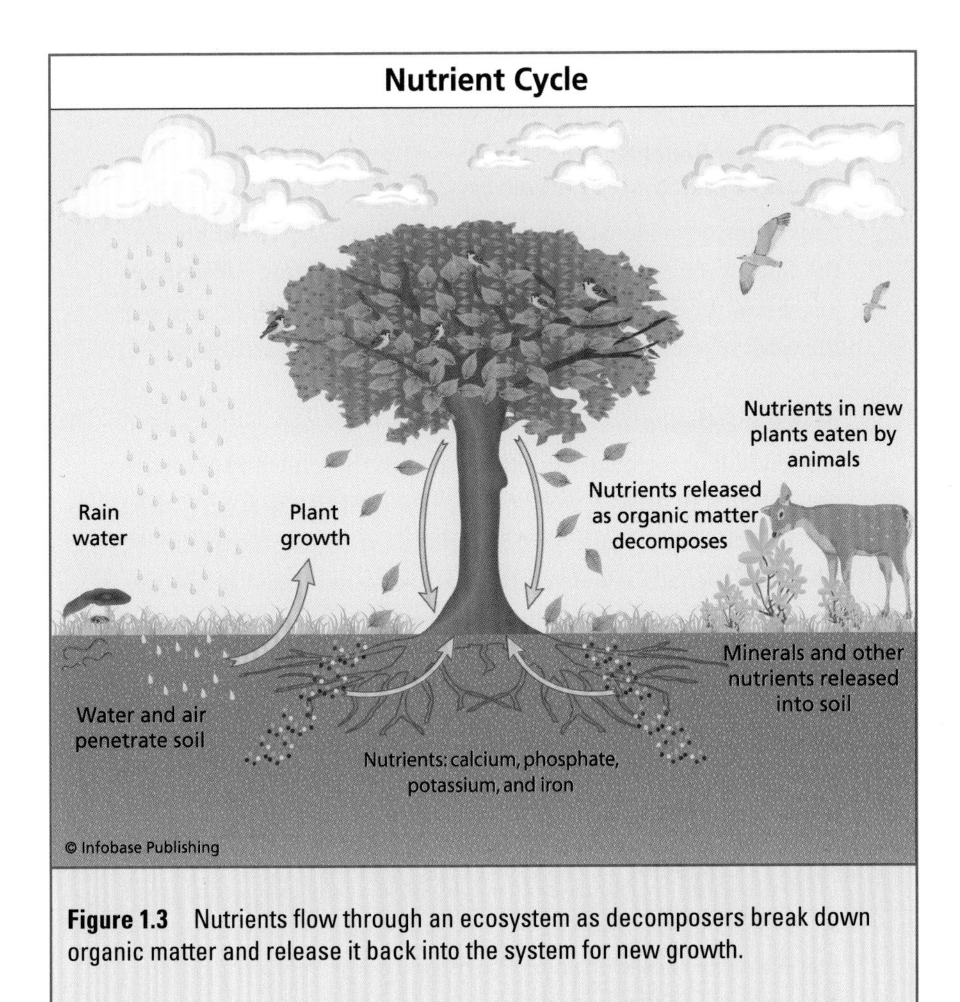

Figure 1.3 Nutrients flow through an ecosystem as decomposers break down organic matter and release it back into the system for new growth.

What Is Biodiversity and Why Is It Important?

The concept of **biodiversity** was introduced in the early 1990s. Pulitzer Prize-winning author and Harvard professor E.O. Wilson gave people a new way to think about why a diversity of species, or biodiversity, is important to a healthy planet.

As Wilson explained, biodiversity, short for biological diversity, organizes life on three levels: the ecosystems (for example, rain forests or coral reefs); the species (the organisms in the ecosystem—humans or red maple trees or mallard ducks are examples); and the variety of genes that are responsible for heredity, which determines the individual nature of each organism.[12]

Why Is Maintaining Biodiversity Important?

Ecological Benefits

Species play an important role in keeping the ecosystems healthy. Ecosystems provide us with amazing benefits: the air we breathe, the clean water we drink, and the soil we use to grow our food are just a few. As E.O. Wilson stated, the ecosystems that sustain our life work so well because the organisms in them are so diverse.[13]

Medical Benefits

Almost half of the pharmaceuticals used in the United States today were originally derived from plants. One example is taxol, a drug derived from the Pacific yew tree and used for cancer treatment. The drug digitalis is derived from a compound found in the foxglove flower (Figure 1.4). Animal compounds can provide important medicines also. For example, calcitonin, a hormone used for the treatment of osteoporosis, and protamine sulphate, an important medicine used in open-heart surgery, both come from salmon.[14]

Agricultural Benefits

About 200,000 wild plant species exist, of which approximately 1,000 are eaten by humans. However, 80% of the crops consumed by the world come from only 12 species of plant.[15] One concern is that if disease should strike primary food crops, replacement crops

Figure 1.4 Foxglove *(Digitalis purpurea)* is the source for digitalis, a drug prescribed by doctors to strengthen the heart and regulate its beat. The name foxglove is derived from its bell-shaped, tubular leaves that resemble the fingers of a glove.

may have already disappeared from Earth. Also important to agriculture are the insects, birds, bats, and other animals that each year pollinate crops worth billions of dollars.

Economic Benefits

From the $28 billion annual income from recreation in national parks and other public lands to the billions of dollars' profit from agriculture, humans benefit financially from a diversity of life on the planet. Consider the economic benefits of coral reefs, one of the most ecologically diverse ecosystems on the planet. Coral reefs provide a tourist industry worth billions of dollars a year in the Florida Keys alone.[16] In addition, the coral reef resources in the United States and its territories contribute an estimated $375 billion to the U.S. economy every year.[17]

A study published in *Nature* projected the economic value of one hectare of land in the Peruvian Amazon rain forest to be about $7,000 annually if harvested using sustainable practices and $148 a year if cleared for cattle pasture.[18]

Right to Exist

Some believe life on the planet has a right to exist, whether we use it or not. The biologist Rachel Carson, in accepting the Albert Schweitzer Medal of the Animal Welfare Institute in 1963 for her book *Silent Spring*, discussed Schweitzer's idea of "Reverence for Life." In her acceptance speech, she said of the concept:

> Whatever it may be, it is something that takes us out of ourselves, that makes us aware of other life. Dr. Schweitzer has told us that we are not being truly civilized if we concern ourselves only with the relation of man to man. What is important is the relation of man to all life... By acquiescing in needless destruction and suffering, our stature as human being is diminished.[19]

Knowledge to Be Gained

About 1.5 million species have been identified, yet, the numbers continue to grow. For example, between 1985 and 2001, the number of identified amphibian species grew from 4,003 to 5,282 species.[20] Consider that two-thirds of the planet is covered by ocean, and two-thirds of the ocean is more than 2 miles (3.2 kilometers) deep. The deep ocean, most of which is unexplored, is the predominant habitat for living things on the planet.[21] Destruction of species before we even learn of their value in the ecosystem or to humans is a concern to many.

As the conservationist Aldo Leopold wrote in 1949:

> The last word in ignorance is the man who says of an animal or plant: "What good is it?" If a land mechanism as a whole is good, then every part is good, whether we understand it or not....To keep every cog and wheel is the first precaution of intelligent tinkering.[22]

In addition, understanding and protecting the diversity of our planet's life also has the potential to answer questions about evolution and life on other planets.

What Factors Cause Species to Become Endangered?

The 2004 Red List developed by the IUCN estimates that almost one in four of Earth's mammal species and one in eight of the bird species are at some degree of risk.[23] As E.O. Wilson pointed out, we are no longer looking at extinctions that happened in the past, but we now have to look at possible extinctions in the future.[24] To deal with the present problem and to plan for the future, one must recognize and consider the broad range of threats to biodiversity.

Habitat Loss

The greatest threat to most species is the loss of habitat, the place that provides food, protection, and a place to reproduce. Habitat is lost when land is cleared for development or logging; when areas are "fragmented" or cut into smaller pieces by roads, thereby reducing a species' territory or range; or when the quality of the habitat is degraded. Another form of habitat loss occurs when humans change river flows by building dams or levees and diverting water for agriculture or domestic needs. This assault on river habitats has reached epic proportions.[25]

Habitat loss is of great concern in the United States and around the globe. In the United States alone, more than half of the **wetlands** have been destroyed[26] and about 90% of the old-growth forests are gone.[27] The tropical rainforests, which support more than 50% of all types of species on the planet, are disappearing at a rate of 214,000 acres (86,000 hectares) a day, which is 78 million acres (31 million ha) per year.[28] As the world population continues to increase, the human need for land and water continues to grow. To date, meeting these needs has meant a continued loss of habitat for species already dependent on those land and water resources.

Population Growth

More than 6 billion people live on the planet today. The United Nations Population Division estimates that by the year 2050, that number could be 13 billion people.[29] The growing population and the associated land-use changes and increase in resource consumption (for example, water, oil, coal, and trees) are considered major factors threatening biodiversity today.[30]

Pollution

In the 1960s, Rachel Carson brought the direct threats of pollution to national attention with her best-selling book, *Silent Spring*. Scientists continue to document the fact that pollution, including pesticides, oil spills, and runoff from agriculture and roads, contribute to the loss of biodiversity.[31] The Pews Oceans Commission 2003 report states that today our oceans, which include open-ocean, deep-ocean, coastal, and coral-reef ecosystems, are facing as great a danger from pollution as our rivers and lakes did in the early 1970s, when polluted rivers caught fire.[32]

Overharvesting

Experience has shown that plants and animals can be hunted or harvested to near extinction: the passenger pigeon, the Steller's sea cow, and the great auk are but a few examples. Increases in the killing of monkeys and apes by local people for bushmeat, or meat to sell in cities, is another. Humans have shown that oceans do not hold unlimited supplies of animals to be harvested. Some whale species have been hunted to near extinction. Various fishing industries, including those in New England and Nova Scotia, have

crashed because of overharvesting that has resulted from industrial fishing. Some estimates put the by-catch (unintentional catch) waste at 25 % of the catch.[33] According to a 2003 report by the Pews Ocean Commission to Congress:

> Overfishing, destructive fishing practices and other threats to fish populations exact a heavy toll. Thirty percent of assessed populations are fished unsustainably, with an ever-growing number of species on their way to extinction as a result.[34]

Climate Change

On the basis of years of observation, scientists are predicting that increased global warming will lead to sea-level rises, disturbances in patterns of rainfall and regional weather patterns, and changes in plants' and animals ranges' and reproductive cycles.[35,36]

A report by the International Climate Change Partnership showed a connection between climate change caused by humans and other environmental issues, including loss of biological diversity (plant and animal species). The opinion of more than 2,000 international experts is that a strong connection exists between global warming and the potential for damage to our planet.[37]

Invasive or Introduced Species

Because ecosystems have developed relationships over millions of years, the sudden introduction of a new species can trigger unnatural changes. In almost 70% of the cases in which native species are driven to extinction, introduced species are a major factor.[38] One example is the millions of dollars in damage done to surface waters because of the introduction of zebra mussels (Figure 1.5), spread when ship ballast water from the Caspian Sea was emptied into the Great Lakes.[39] Some plant or animal species are introduced after they have been imported by people and then are released or spread. Whatever the source, invasive species are a serious threat to ecosystems.

The list of serious threats is growing, but what is being done to protect endangered species?

Figure 1.5 Zebra mussels *(above)*, native to Eastern Europe and Western Asia, are considered to be a type of invasive aquatic species. In 1988, the mussels were first found in the Great Lakes after being transported by ballast water from the Caspian Sea. The zebra mussel is now spreading rapidly and clogging power plant water supply systems.

HISTORY OF THE ISSUE

Programs to protect endangered species and their habitats have existed in a variety of forms for more than 100 years. Regulations and policies at both international and national levels have been in place for more than 50 years. In addition, a variety of organizations and programs around the world have developed over the past 100 years to protect species or the habitats in which they live. The following section highlights some notable historic developments.

What Legislation and Policies Protect Endangered Species?

International Whaling Commission

More than 50 years ago, the international community recognized that more than a century of whaling had left much of the whale population threatened with or near extinction. In 1946, the International Whaling Commission (IWC) was established: "Having decided to conclude a convention to provide for the proper conservation of whale stocks and thus make possible the orderly development of the whaling industry."[40]

Membership in the IWC is open to any country willing to comply with the recommendations of the 1946 agreement. In 1986, the IWC declared a moratorium on commercial whaling until the whale populations recovered. A Revised Management Scheme (RMS) was accepted in 1994 but has not yet been implemented.

Today the IWC has 62 members. Norway, Iceland, and Japan, although signatories, continue to hunt whales either by using a loophole in the agreement that allows hunting for scientific purposes or by simply disregarding the agreement.[41]

CITES and TRAFFIC

Endangered elephant and tiger populations have been in the news for decades. In the 1960s, many countries realized that addressing the problems of trading and exploiting wild animals and plants had to be done on an international level. Countries began talking about protection in the early 1960s, and finally, in 1975, 30 countries agreed to CITES (pronounced *site-ease*), or the Convention on International Trade in Endangered Species of Wild Fauna and Flora.[42] Today, CITES has grown to a voluntary membership of 167 governments.[43] Its purpose is to regulate the trade in specimens of wild plants and animals in a way that protects them from becoming endangered.

TRAFFIC, the joint wildlife-trade monitoring program of the World Wide Fund for Nature and the World Conservation Union, was founded to assist CITES member countries in implementation of the program. Since the 1970s, it has grown to an organization with 22 offices around the world. In addition to regional issues, TRAFFIC also addresses commercial trading, such as fisheries and timber.[44]

United States National Wildlife Refuge System

A century ago, President Theodore Roosevelt brought strong conservation leadership to our country. One gift he gave to our nation was the establishment of the United States National Wildlife Refuge System, with the specific purpose of protecting the land on which plants and animals live. The first refuge was on Pelican Island, Florida (Figure 1.6), designated in 1905 to offer refuge to the many shorebirds that were being slaughtered for their feathers, which were used to decorate ladies' hats.

A system that began as a single, 5.5-acre (2.2 hectare) refuge on Pelican Island has grown to include 540 refuges, with at least one in every state. The refuges of the United States cover 95 million acres of diverse habitats that include wetlands, water, desert, forests, and coastlines. On these protected lands and waters live almost 300 species of endangered or threatened vertebrate animals.[45]

Marine Protected Areas

In 1972, 100 years after Yellowstone National Park was established, the United States passed the National Marine Sanctuary Act. Its purpose was to protect marine resources, while managing them for compatible uses such as fishing, diving, and certain commercial activities. Part of the challenge to implementation of the act is to balance resource protection with economic growth. Since 1972, the program has expanded to encompass 13 marine sanctuaries that include such diverse ecosystems as coral reefs, deep-sea canyons, historic sites, and deep oceans. These sanctuaries protect an area nearly the size of Vermont and New Hampshire combined, or 18,000 square miles (46,619.8 square kilometers).[46]

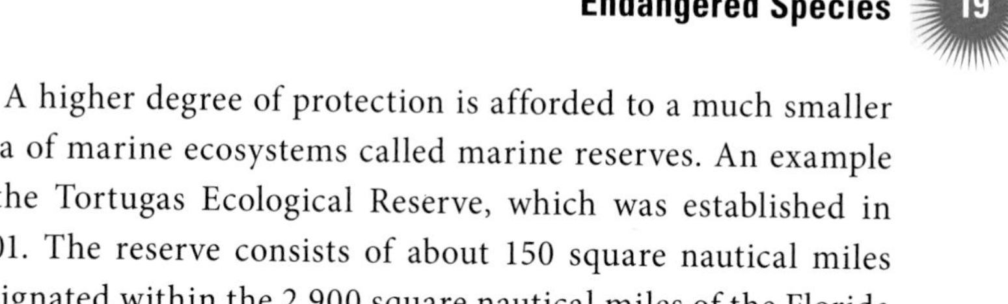

A higher degree of protection is afforded to a much smaller area of marine ecosystems called marine reserves. An example is the Tortugas Ecological Reserve, which was established in 2001. The reserve consists of about 150 square nautical miles designated within the 2,900 square nautical miles of the Florida Keys Sanctuary. After seeing the economic and environmental benefits of the reserve, 78% of area residents supported the extra protection offered by reserves.[47] The protection afforded to marine areas, both sanctuaries and reserves, is an important tool in protecting threatened and endangered species.

Figure 1.6 The Pelican Island National Wildlife Refuge *(above)* is America's first national wildlife refuge. More than 30 species of birds nest on Pelican Island, as well as threatened and endangered species such as the Florida manatee and loggerhead sea turtles.

United States Endangered Species Act

In 1973, President Nixon reminded our country that "Nothing is more priceless and more worthy of preservation than the rich array of animal life with which our country has been blessed."[48] The occasion was the updating of an existing, less-stringent, 1966 law to protect endangered species. The purpose of the new act, called the Endangered Species Act (ESA) was to conserve endangered and threatened species and the ecosystems on which they depend for life.[49] The act offered a legal expression for what ecologist G.E. Hutchinson once wrote: "Species are the actors in the ecosystem theater."[48]

As science offers more understanding about the way in which ecosystems work, the act continues to be amended. Section 4 of the ESA states that the FWS must develop a recovery plan for species listed as endangered or threatened. Today, more than 1,200 species of plants and animals in the United States are listed as endangered or threatened. About 400 of these species are still in decline, whereas about a third of the species are stable or improving. [49] The red wolf, the black-footed ferret, whooping cranes, the bald eagle, and the California condor are examples of species that have recovered under the ESA.

Clean Water Act

In response to public outcry over the pollution of rivers and lakes, Congress passed the Federal Water Pollution Control Act in 1972. The act was then amended in 1977 and became known as the Clean Water Act (CWA). The significance of the act was that it gave the Environmental Protection Agency (EPA) authority to implement pollution-control standards. Until the passage of the CWA, the dominant policy in this country was to drain and fill wetlands. Section 404 of the CWA is the primary protection afforded to wetlands in this country. This protection was significant because almost half of the animal species listed as endangered in 1986 were dependent on wetlands.[50]

National Invasive Species Act

The National Invasive Species Act (NISA) of 1996 strengthened the Nonindigenous Aquatic Nuisance Prevention and Control Act of 1990. In April of 2005, NISA was presented to Congress for reauthorization. The act established a task force of various government agencies to implement programs and to do research on invasive species.[51]

What Are Some Organizations That Protect Endangered Species?

International organizations, such as the World Conservation Union, bring together governments and other nongovernmental organizations (NGOs) to protect endangered species and their habitats. Various nongovernmental organizations work on the local, national, and international level through programs that include land-preservations projects, and they work with local governments and communities in economic-development programs to protect species and habitats. Examples of NGOs include the National Audubon Society, the World Wildlife Fund, Conservation International, the Wildlife Conservation Society, and The Nature Conservancy.

The World Conservation Union

The World Conservation Union, or IUCN, was founded in 1948. Its mission is "to influence, encourage and assist societies throughout the world to conserve the integrity and diversity of nature and to ensure that any use of natural resources is equitable and ecologically sustainable."[52] Since 1948, the IUCN has been developing databases and reports to be used in formulating policies and plans to protect natural resources. Today more than 10,000 scientists and experts contribute to IUCN reports. A major report generated by the IUCN is the Red List of Threatened Species (www.redlist.org). Information in the Red List helps shape global policy on protecting biodiversity.

World Wildlife Fund

The World Wildlife Fund (WWF) was established in 1962 as an international organization that had the scientific, financial, and technical resources to develop conservation projects around the world. Its primary mission is the protection of endangered species and their habitats. Over the years, the WWF has expanded its approach to solutions to include policy, education, economics, and science. The WWF works in more than 100 countries to find economically and environmentally sustainable solutions to preserve biodiversity.[53]

The WWF was the major force behind the international ban on ivory trading in 1990. It has also worked to create more than 500 parks in Africa, Latin America, and Asia. In the past 10 years, it has worked to protect more than 1 billion acres of forest habitat around the world.[54] The WWF also does important conservation work in the United States.

The Nature Conservancy

The Nature Conservancy (TNC) started in 1951 as an outgrowth of the Ecological Society of America, which was formed in 1915. The mission of The Nature Conservancy is "to preserve the plants, animals and natural communities that represent the diversity of life on Earth by protecting the lands and waters they need to survive."[55] A key protection tool used by TNC is land acquisition and protection. Today TNC has protected more than 117 million acres (473,482 sq km) of land and 5,000 miles (8,046 km) of rivers around the world. TNC works closely with other conservation groups, corporations, native peoples, and governments to accomplish its environmental goals.[56]

Other Organizations

Other groups, such as Environmental Defense (ED), National Wildlife Federation (NWF), Natural Resources Defense Council (NRDC), and the Sierra Club, have taken a different approach to protecting our environment. In addition to educating the public on environmental issues, these groups serve as legislative watchdogs by monitoring industry and the government, while educating citizens to rally support for environmentally responsible behavior. Environmental Defense has, since 1967, "linked science, economics and law to create innovative, equitable and cost-effective solutions to society's most urgent environmental problems."[57]

CURRENT ISSUES AND FUTURE CONSIDERATIONS

How Many Species Have Gone Extinct or Are Listed As Threatened or Endangered?

Background extinction rate refers to natural extinction rates that cannot be attributed to humans. The natural rate of extinction is considered by experts to be one species per million per year.[58]

Today's rates are estimated to be as much as 1,000 times higher than the background rate. Nearly one in four mammal species is in serious decline because of human activity.[59] In the United States, we have at least 90 different species of frogs and toads and 140 species of salamanders. Today, 27 species of amphibians are listed as endangered or threatened.[60] Over the past 200 years, more than 100 species of birds have gone extinct.[61] According to the IUCN Red List, 12% of today's birds are endangered or threatened.[62] Several populations of whales remain "highly endangered," with total numbers of 500 or less. The North Atlantic right whale, the Western North Pacific gray whale, and the blue whale are examples.[63] However, the extinctions are not happening only to animals. More than 8,000 species of plants around globe are threatened with extinction.[64] E.O. Wilson has predicted that if humans keep altering Earth, we could lose a fifth or more of our plant and animal biodiversity by 2020.[65]

What Are the Policy Issues?

Endangered Species Act

Since its enactment in 1973, the Endangered Species Act has been criticized by some for not providing enough protection to endangered or threatened species. They offer the low numbers of plants and animals that have been "delisted," or taken off the endangered list, as proof. Yet, others argue that in addition to inhibiting economic growth and depriving landowners of rights, the ESA has failed to protect species—also citing the same fact that only 1% of the plants or animals have recovered enough to be taken off the endangered species list. Protecting critical habitat and balancing economic concerns are issues that will continue to keep endangered species in the news.

The ESA is currently being reviewed by Congress. Senator Hillary Rodham Clinton, a member of the committee reviewing the ESA, has said, "We are on the brink, according to scientists, of seeing many forms of life disappear. This is not as dire in the United States as elsewhere, and I would argue strongly that one reason is the ESA. We ought to be proud."[66]

Fishing Regulations

Just one look at a map of the world makes clear that the oceans are truly a shared global resource. Therefore, the fishing policies of one country or area have effects that reach beyond any national fishing boundaries.

The International Whaling Commission (IWC) issued a moratorium on whaling in 1986 that is still in place—yet, compliance is voluntary. In the United States, the 1996 Sustainable Fisheries Act brought the new language of "sustainable yield" into fishing laws. Now, fisheries have to set catch limits that allow a particular fish species to maintain itself.[67]

Since the 1970s, the European Union has regulated fishing in Europe through the common fisheries policy (CFP). Their goal is to conserve fish stocks, protect the environment, and ensure the economic viability of the European fleets.[68]

The 2003 Pews Ocean Commission Report to Congress recommends that a "unifying vision of ocean stewardship" is needed in this country to consolidate existing federal agencies and that these agencies be directed to protect and to restore ocean resources.[69]

Clean Water Act

Passage of the Clean Water Act in 1972 was an important step in protecting the habitats of endangered species. However, the Clean Water Act continues to be amended, with some changes and policy interpretations potentially weakening the intended protections. Future considerations are keeping the legal framework for protecting water quality and wetlands in place so that habitats are not degraded or destroyed.

What Are Some Management Strategies?

Hot Spots

Some areas of the world are losing species faster than others. Also, some ecosystems are much richer in numbers of different species than others. Coral reefs are the most diverse ecosystem in the oceans—sometimes they are called the rain forests of the ocean. On land, tropical rain forests cover only 7% of Earth's land surface but contain more than half the species in the entire world. Tropical rain forests, among the most fragile ecosystems on Earth, are currently being destroyed at such an alarming rate that they will almost entirely disappear in the next 100 years, taking hundreds of thousands of species into extinction. Other species-rich ecosystems are coastal wetlands.[70]

These facts went into developing the concept of "hot spots," first written about in a 1988 paper by Norman Myers. To qualify as a hot spot, an area must have at least 1,500 different kinds of plants, and it has to have lost at least 70% of its original habitat. An updated analysis by Conservation International has targeted 34 hot spots around the world. Together these areas hold more than 50% of the world's endemic, or native, plant species. They also hold 77% of the world's total of land vertebrates (animals with backbones).[71]

Protected Areas

Today, the World Commission on Protected Area, part of the IUCN, is the leading international network of protected-area specialists. The World Conservation Congress defines a protected area as an area of land, sea, or both especially dedicated to the protection and maintenance of biological diversity and of natural and associated cultural resources, managed through legal or other effective means.[72]

Today, 100,000 protected areas cover about 11% of the surface of the globe. However, many biologically rich habitats are not included in protected areas, and some protected areas need better management. Future challenges include addressing key ecosystems and connecting protected areas to social and economic concerns. Identifying these priority areas helps conservation groups target them for protection efforts, including working with local people, governments, and other organizations.

Figure 1.7 A flagship species is a species designated as the logo for a specific environmental cause. The World Wildlife Fund, a conservation organization, uses the giant panda as their flagship species.

Flagship Species

Most people do not get excited about saving snails or frogs. So the idea of designating flagship species, those with more popular appeal, was developed to rally support to protect species and their habitats. An example is the World Wildlife Fund logo, the giant panda (Figure 1.7).

When one large species and its habitat are protected, many other species in that habitat also benefit. Flagship species, including tigers, rhinos, whales, sea turtles, great apes, and elephants, in addition to giant pandas, are species used by the World Wildlife Fund to target protection strategies. Other organizations have used similar protection strategies.

Partnerships

Combining efforts and resources to protect endangered species and the ecosystems in which they live allows organizations and governments to tackle larger problems than any of them could take on alone. Many nongovernmental groups (for example, the World Wildlife Fund, The Nature Conservancy, and Conservation International) form partnerships with other conservation groups and with businesses, industries, and all levels of governments. Sharing financial resources, political power, and scientific expertise is proving an effective way to strengthen conservation efforts around the world.

WHAT THE INDIVIDUAL CAN DO

1. When you travel, do not buy products made from any endangered animal parts. These products include skins, body parts, tortoise shells, ivory, and coral. Any of these could be from endangered species, and by purchasing them, you might be breaking the law.
2. Do not let nonnative species you keep as pets (such as snakes, lizards, or insects) loose in the wild.

For example, bullfrog pollywogs grow into bullfrogs, which can cause much damage to ponds and lakes.

3. Keep water and wetlands clean. Do not flush oils, paints, degreasers, and other harmful chemicals down your sink. The water eventually ends up somewhere—in groundwater or in rivers or lakes.
4. Conserve energy. The more energy we save, the less carbon dioxide gets released into the atmosphere, slowing global warming.

REFERENCES

1. Hoose, Philip. *The Race to Save the Lord God Bird.* New York: Farrar, Straus, and Giroux, 2004.
2. "The endangered species listing program," U.S. Fish and Wildlife. Available online. URL: http://endangered.fws.gov/listing/index.html.
3. Tuxill, J. and C. Bright. "Losing strands in the web of life." In *State of the World.* New York: Norton, 1998.
4. Wilson, E.O. *The Diversity of Life.* Cambridge, Mass.: Harvard University Press, 1992.
5. Wilson, E.O. *The Future of Life.* New York: Vintage Books, 2002.
6. Johnson, George B. *The Living World.* New York: McGraw Hill, 2003.
7. Starr, Cecie, and Ralph Taggart. *Biology: The Unity and Diversity of Life.* Belmont, Calif.: Wadsworth Publishing, 1998.
8. Ray, Justina C., et al. *Large Carnivores and the Conservation of Biodiversity.* Washington, D.C.: Island Press, 2005.
9. Soule, M.E., J.A. Estes, J. Berger, and C. Martinez del Rio. "Ecological effectiveness: Conservation goals for interactive species." *Conservation Biology,* Volume 17, October 2003: 1238–1250.
10. Estes, James A. "Carnivory and trophic connectivity in kelp forests." In Ray, Justina C., et al. *Large Carnivores and the Conservation of Biodiversity.* Washington, D.C.: Island Press, 2005, p. 80.
11. Muir, John. *My First Summer in the Sierra.* Boston: Houghton Mifflin, 1911 (Sierra Club Books edition, 1988).
12. Wilson, *Diversity of Life,* p. 51.
13. Ibid., p. 347.
14. World Wildlife Fund. "Factsheet: Biodiversity," Available online. URL: http://www.biodiversity911.org/biodiversity_basics/why_important/DiversePrescription.html. 2005.
15. Diamond, Jared. *Guns, Germs, and Steel: The Fates of Human Societies.* New York: Norton, 1997.
16. "Healthy coral reefs provide income: Billions of dollars for U.S. economy, millions of jobs," NOAA Coral Reef Conservation Program. Available online. URL: http://www.coralreef.noaa.gov/. 2004.

17. NOAA. "Marine sanctuaries factsheets," Available online. URL: http://www.sanctuaries.nos.noaa.gov/. 2004.
18. Peters C.M., A.H. Gentry, and R.O. Mendelsohn 1989. "Valuation of an Amazonian rainforest." *Nature* 339 (June 29, 1989): 655–656.
19. Brooks, Paul. *The House of Life: Rachel Carson at Work*. Boston: Houghton-Mifflin, 1972.
20. Wilson, *Diversity of Life*.
21. Ellis, Richard. *The Empty Ocean: Plundering the World's Marine Resources*. Washington, D.C.: Island Press, 2003.
22. Leopold, Aldo. *A Sand County Almanac*. New York: Oxford University Press, 1949.
23. "The IUCN Red List of threatened species: Summary statistics for globally threatened species," Available online. URL: http://www.redlist.org/info/tables.html. 2004.
24. Wilson, *The Future of Life*, p. 100.
25. Postel, Sandra and Brian Richter. *Rivers for Life: Managing Water for People and Nature*. Washington, D.C.: Island Press, 2003.
26. Mitsch, William J. and James G. Gosselink. *Wetlands*. 3rd Edition. New York: John Wiley and Sons, 2000.
27. Starr, *Biology: The Unity and Diversity of Life*, p. 488.
28. "Rates of rainforest loss," Rainforest Action Network. Available online. URL: http://www.ran.org/info_center/factsheets/04b.html. 2005.
29. The Worldwatch Institute. *State of the World 2004*. New York: Norton, 2004.
30. Spray, Sharon. L. and Karen L. McGlothlin, eds. *Loss of Biodiversity*. Lanham, Md.: Rowman & Littlefield, 2003.
31. Ibid.
32. Pew Oceans Commission. "America's living oceans: Charting a course for sea change." Available online. URL: http://www.pewtrusts.org/pdf/env_oceans_species.pdf. 2003.
33. Ellis, *The Empty Ocean*, p. 16.
34. Pews Oceans Commission. "America's living oceans: Charting a course for sea change." Available online. URL: http://www.pewtrusts.org/pdf/env_oceans_species.pdf. 2003.

35. Intergovernmental Panel on Climate Change. "Technical report V: Climate change and biodiversity." Available online. URL: http://www.grida.no/climate/ipcc_tar/. 2001.
36. Semlitsch, Raymond D. (ed). *Amphibian Conservation.* Washington DC.: Smithsonian Books, 2003.
37. Intergovernmental Panel on Climate Change. IPCC Third Assessment Report. "Third assessment report: Climate change 2001: Synthesis Report: Summary for Policymakers," IPCC. Available online. URL: http://www.ipcc.ch/pub/un/syreng/spm.pdf. 2001.
38. Starr, *Biology*, p. 489
39. Spray, *Loss of Biodiversity*, p. 53.
40. International Whaling Commission. "International convention for the regulation of whaling, Washington, D.C. Dec 2, 1946," Available online. URL: http://www.iwcoffice.org/commission/convention.htm.
41. International Whaling Commission. Information. Available online. URL: http://www.iwcoffice.org/commission/iwcmain.htm#nations. 2005.
42. World Wildlife Fund. "Factsheet: Wildlife trade," Available online. URL: http://worldwildlife.org/trade/cites/about.cfm.
43. CITES. "Secretariat factsheet," Available online. URL: http://www.cites.org/eng/disc/what.shtml. 2005.
44. TRAFFIC. "Factsheet," TRAFFIC International. Available online. URL: http://www.traffic.org/about/. 2003.
45. Graham, Frank Jr. "Safe havens," *Audubon*, 2003: 43.
46. "Welcome to marine sanctuaries factsheet," Available online. URL: http://www.sanctuaries.nos.noaa.gov/oms.oms.html. 2004.
47. Klingener, Nancy. "Providing sanctuary," *Blue Planet*, Fall 2002.
48. The National Oceanic and Atmospheric Administration (NOAA) Fisheries. "Endangered Species Act of 1973 factsheet," Office of Protected Resources. Available online. URL: http://www.nmfs.noaa.gov/prot_res/laws/ESA/ESA_Home.html.
49. United States Fish and Wildlife Service. Endangered Species. Bulletin. Available online. URL: http://www.fws.gov/endangered/esb/2003/07-12/toc.html. 2003: 10.
50. Mitsch, *Wetlands*, pp. 583, 642.

51. Union of Concerned Scientists Factsheet. "The national aquatic invasive species act," Available online. URL: http://www.ucsusa.org/global_environment/invasive_species/page.cfm?pageID=882. 2005.
52. IUCN Factsheet. "About IUCN," Available online. URL: http://iucn.org/about/index.htm. 2004.
53. World Wildlife Fund Factsheet. "About WWF: History," Available online. URL: http://www.worldwildlife.org/about/index.cfm. 2005.
54. Ibid.
55. The Nature Conservancy Factsheet. "About us," Available online. URL: http://nature.org/aboutus/. 2005.
56. Ibid.
57. Environmental Defense Factsheet. "About Environmental Defense," Available online. URL: http://www.environmentaldefense.org/aboutus.cfm?subnav=aboutus. 2005.
58. Stein, Bruce, A., et al, eds. *Precious Heritage: The Status of Biodiversity in the United States*. New York: Oxford University Press, 2000.
59. Worldwatch Institute. *Vital Signs 2005*. New York: Norton, 2005.
60. Semlitsch, *Amphibian Conservation*, p. 3.
61. Youth, Howard. *Winged Messengers: The Decline of Birds*. Washington, D.C.: The Worldwatch Institute, 2003.
62. IUCN. "Red List," Available online. URL: http://www.redlist.org/. 2004.
63. International Whaling Commission. "The International Whaling Commission's 56th annual meeting." Press Release, 2004.
64. IUCN. "Red List," Available online. URL: http://www.redlist.org/. 2004.
65. Wilson. *The Future of Life*.
66. Today's News. "Defenders of wildlife," Available online. URL: http://defenders.org/newsroom/whatsnew.htm. June 8, 2005.
67. Safina, Carl. "The continued dangers of overfishing." *Issues in Science and Technology*, Summer, 2003.
68. Fisheries and Maritime Affairs. "*Factsheet 2.2: The common fisheries policy*," Available online. URL: http://europa.eu.int/comm/fisheries/reform/index_en.htm. 1998.
69. Pews Ocean Commission 2003.

70. Wilson, E.O., ed. *Biodiversity. Washington D.C.: National Forum on Biodiversity.* Washington D.C.: National Academies Press, 1988.
71. Conservation International. "*Hot spot science factsheet*," Available online. URL: http://www.biodiversityhotspots.org/xp/Hotspots/hotspotsScience/key_findings. 2005.
72. IUCN Factsheet. "Welcome to the IUCN World Commission on protected areas," Available online. URL: http://www.iucn.org/themes/wcpa/. 2005.

FURTHER READING

Books

Wilson, E.O. *The Diversity of Life*. New York: Norton, 1999.

Wilson, E.O. *The Future of Life*. New York: Time Warner Books, 2003.

Ray, Justina C., et al, eds. *Large Carnivores and the Conservation of Biodiversity*. Washington, D.C.: Island Press, 2005.

Clark, Tim W., et al, eds. *Coexisting with Large Carnivores: Lessons from Greater Yellowstone*. Washington, D.C.: Island Press, 2005.

Web Sites

Conservation International
http://www.conservation.org

Convention on Biological Diversity
http://www.biodiv.org/programmes

The Nature Conservancy
http://www.nature.org

Red List of Endangered Species
http://www.iucnredlist.org

U.S. Fish and Wildlife Service
http://www.fws.gov

World Wildlife Fund
http://www.worldwildlife.org

Air Pollution

WHAT'S IN THE NEWS?

Acid rain and unhealthy levels of air pollution are in the news. For example, pollution from coal-fired plants in the Midwest is causing soccer games to be cancelled in Vermont and falling as acid rain in New York lakes. The Environmental Protection Agency (EPA) tells us that more than 150 million Americans live in areas with unhealthy levels of air pollution (Figure 2.1). In April of 2004, the EPA announced that 474 counties across the United States were out of compliance with smog standards.[1]

The United States began thinking seriously about air pollution in the 1950s, when Congress allocated a small amount of money to study the problem. Since that time, great advances in science have given policy makers the scientific facts needed to fashion comprehensive regulations to protect human health and the environment. These protections turned out to have major economic benefits as well. Stricter regulations have helped clean the air, even as America's Gross National Product (GNP) continues to grow.

Although great improvements in air quality indicate the effective strategy of regulating pollution sources, the job is not yet done. Nitrogen oxides and sulfur dioxides from power plants still pose major air-quality threats to people and to our environment. The problem is not just a local problem. Many communities with unhealthy levels of air pollution receive more than 30% of the pollution from upwind states.[2] The EPA estimates that about 75% of emissions from U.S. coal plants are blown outside of America's borders.[3] Automobile pollution still chokes some large cities with mucky haze. Billions of dollars in health costs and environmental damage from air pollution tells us that air quality will continue to be in the news for years to come. Hopefully, as in the case of the 98% reduction in lead pollution in the United States, the news will improve.

Figure 2.1 The above photo of the Utah Power & Light generating plant depicts the way in which the air is contaminated with industrial pollutants. Air pollution is a leading cause of ozone depletion and respiratory disease.

THE SCIENCE

What Is the Atmosphere?

Air, like water, moves freely over the planet and does not recognize national boundaries. Circulating air moves pollution as air currents travel far and wide. Earth's atmosphere consists of gas and particulate components and the physical forces that act upon and within them.[4] Gravity holds atmospheric gases near Earth's surface.

The most abundant gases are nitrogen (N_2), which makes up about 78% of the atmosphere, and oxygen (O_2), which makes up about 21% of the atmosphere. Trace gases, such as carbon dioxide, carbon monoxide, methane, and nitrous oxides, along with water vapor, make up the remainder. Particulates include dust, **aerosols**, and other solid particles suspended in air.

Air currents, like water currents, are powered by solar energy. Because the Earth is round, the solar heat hits Earth at different angles, with more energy received near the equator than near the poles. Warmer air currents naturally move toward the cooler polar areas. Air also moves as energy is transferred in the process of **evaporation** and **condensation**.[5] How fast can air currents move? The jet stream can reach velocities of 400 km (248 mi) per hour.[6]

The atmosphere is layered according to temperature differences. The troposphere is the lower layer, extending above Earth's surface from about 10 miles (16 km) in the tropics to 5 miles (8 km) in the higher latitudes. In this layer, the air masses mix because of differences in temperatures. The next layer, the stratosphere, lies up to 30 miles (50 km) above Earth's surface. Little mixing and no precipitation occurs in this layer.

What Are Air Pollutants?

Natural

Volcanic eruptions and forest fires degrade air quality, filling the air with particulates and smoke. Wind erodes soils and fills the air with dust particles. Plants give off volatile **hydrocarbons**. These examples are a few of the types of natural air pollution. Because humans cannot regulate or control these sources, we are fortunate they do not pose any sort of serious long-term threat. Pollution generated by natural sources is spread out over the globe, not concentrated in populated areas, and occurs in much smaller amounts than pollution generated by humans. The larger events, such volcanic eruptions or huge forest fires, do not occur with any great frequency.[7]

Anthropogenic

Anthropogenic sources of pollution are those related to human activity. The levels of oxides (e.g., sulfur, carbon, and nitrogen compounds) traceable to human sources have increased fourfold since 1900. The EPA has identified six principal air pollutants and a list of 188 air toxins. The six principal, or "criteria," pollutants for which the EPA has set national air-quality standards include nitrogen oxides (known as NOx), **particulate matter** (known as PM), carbon monoxide (CO), sulfur dioxide (SO_2), **ozone** (O_3), and lead (Pb).[8]

Because scientific data on which to base policy and regulations is needed, air sampling for the EPA's ambient air-quality program is done at the state and local level. Four thousand monitoring sites nationwide collect data on the six principal pollutants and also on ozone precursors, which include approximately 60 volatile hydrocarbons and carbonyls.[9]

Nitrogen oxides

Nitrogen is a relatively stable compound. However, when nitrogen is involved in **combustion** or biological processes, it becomes highly reactive and plays a notable role in atmospheric chemistry.[10] These compounds contribute to ground-level ozone pollution and smog.

The term "nitrogen oxides" describes several nitrogen compounds and is given the notation "NOx." One such compound, nitrous oxide (N_2O), is of concern because is depletes the stratospheric ozone layer and also absorbs heat, thus contributing to global warming. Another nitrogen compound, nitrogen dioxide (NO_2), is produced by a photochemical **oxidation** of nitrogen dioxide and creates a yellow-to-brown irritating gas. The two principal sources for NOx emissions are transportation and stationary-source fuel combustion—the latter source, for example, includes power plants and other industries. Nonroad vehicle sources, such as construction equipment, boats, and airplanes, have increased steadily over the past 10 years.[11]

Sulfur dioxide

Every year, the United State discharges about 30 million tons of sulfur dioxide into the atmosphere. Sulfur dioxide (SO_2) forms when fossil-based fuels, such as coal and oil, are burned. The highest concentrations of SO_2 have been shown around large industrial plants. Power plants generate more than 60% of all the SO_2 produced by human sources.[12] The remainder is emitted from industrial boilers, smelters, and refineries.[13] SO_2 sources are widespread, but higher concentrations are found in the Midwestern United States, where older, coal-burning power plants are not subject to strict emissions controls.

Particulate matter

Particulate matter (PM) refers to dust, soot, aerosols, and smoke particles of various sizes found suspended in the air. It is the black, sooty stuff that spews out of power plants and diesel trucks and ships. It also includes particles formed from sulfur dioxide and nitrogen oxides. PM comes from many sources, including transportation engines, factories, fires, and wind **erosion**. Particles less than 10 micrometers, referred to as PM_{10}, can be inhaled into the body. However, particles under 2.5 micrometers (10 to 20 times smaller than the diameter of a human hair) pose a greater risk because they can become lodged in the lung tissues.

PM and ozone form haze, the microscopic particles or liquid drops in the air. Visibility is obscured because these particles scatter and absorb radiation from the Sun. The fine particles that cause haze include sulfates, nitrates, ammonium, dust, certain metals, some forms of carbon, and salts. Nonroad sources of diesel soot, including commercial marine vessels (ships), farm, construction, and mining equipment, combine to add more pollution than do all the road vehicles.[14]

Another source of PM is one that the United States cannot regulate. Millions of tons of dust are being blown from the African deserts across the Atlantic Ocean every year. The United States Geological Survey (USGS) is tracking the dust clouds and taking samples, concerned that the dust is contributing to the decline of Caribbean coral reefs, causing the increase in asthma, and carrying fungus and bacteria.[15]

Carbon monoxide

Carbon monoxide (CO) is gas with no color or odor. It forms when fuels and biomass are not completely combusted. About 60% of all

Figure 2.2 Due to a weather phenomenon called temperature inversion, Salt Lake City is known to have some of the worst air pollution in the United States. In 2001, Utah's State Capitol building *(above)* is barely visible through the city smog and haze.

CO emissions come from vehicles on the road, and the remainder comes from industrial and other nonroad transportation, such as ships and planes.[16] Concentrations of CO are usually higher during the colder months, when air pollutants become trapped near the ground under a layer of warmer air in what is called an inversion (Figure 2.2).

Ozone

Ozone is a molecule made up of three atoms of oxygen rather than the usual two. When it occurs in the upper layer of the atmosphere, it forms a protective shield from the Sun's ultraviolet radiation. However, when it occurs closer to Earth, it forms a major component of city smog and contributes to many human health problems.

Ozone is formed as a by-product of nitrogen oxides and other volatile organic compounds (VOCs). The chemical reactions are powered by energy from the Sun, so these compounds are called photochemical oxidants. More reactions occur in the heat of the summer, which accounts for higher ozone pollution in some cities during the summer months.[17] Diesel engines, cars, trucks, and industries all emit NOx and VOCs that form ozone and contribute to smog.

Lead

In the 1920s, lead was added to gasoline to improve engine performance. Removal of this harmful lead compound from gasoline took many years to legislate. Lead in the air has traveled to places as far away as Greenland.[18] Today, the primary sources of lead emissions are industrial processes that involve smelting and the manufacture of batteries.[19]

Air toxins

Air toxins are a category of contaminants identified by the EPA that are "known or suspected to cause cancer or other serious health effects, such as reproductive effects or birth defects, or adverse environmental effects."[20] Benzene, dioxin, mercury, asbestos, toluene, cadmium, lead, and chromium are examples of the 188 air toxins from industrial sources listed in the Clean Air Act. Benzene (C_6H_6) is found in fossil-fuel emissions (vehicle exhaust and burning of coal and oil), industrial solvents, and tobacco smoke.

Mercury, another air toxin, is released from chlorine plants, steel plants that recycle automobile parts, and coal-fired power plants, as coal burns. These plants are the largest human source of mercury emissions. Mercury from the air is then deposited in water or on land. Once in water, bacteria change the mercury into a neurotoxin, methylmercury, which then accumulates in the tissues of fish and animals that eat fish—including humans. Scientists collect data on mercury in the environment to gain information to protect human health and the environment.

Also included in the 21 pollutants listed from mobile sources are diesel particulate matter and exhaust organic gases. In addition,

the EPA has identified 33 hazardous pollutants that pose threats to people living in urban areas.[21]

Currently, the EPA compiles an Air Toxics Inventory as part of the National Emission Inventory to monitor the trends for these pollutants. The most recent data available show that the emissions of toxic air pollutants are divided among four sources: (1) large industrial sources; (2) smaller industrial sources such as small dry cleaners and gasoline stations, as well as natural sources such as wildfires; (3) on-road mobile sources, including highway vehicles; and, (4) nonroad mobile sources such as aircraft, locomotives, and construction equipment. Although, on average, the emissions are about equally divided among these sources, the proportions do vary across the country.

What Are the Effects of Air Pollution?

Health

Scientists collect data on toxins and pollutants in the environment to gain information that policy makers can use to protect human health and the environment. Scientific studies over the past three decades have shown that exposure to air pollutants increases the risk of respiratory infections, including pneumonia and bronchitis. Impacts of pollution can be direct, as in lung infections, or indirect, as in toxins absorbed into the bloodstream or heart attacks caused by the stress of chronic respiratory distress.[22] Air pollution is not simply a human health issue. It also affects other living organisms, such as animals and plants.

Given that an average adult breathes about 3,000 gallons (11,356 liters) of air each day and that children breathe even more, the fact that breathing dirty air affects our health is not surprising. For the first time, in May of 2004, the American Heart Association (AHA) released a statement that air pollution is a serious public-health problem:

> During the last decade, however, epidemiological studies conducted worldwide have shown a consistent, increased risk for cardiovascular events, including heart and stroke deaths, in relation

to short- and long-term exposure to present-day concentrations of pollution, especially particulate matter.[23]

Particulate matter and ozone create smog, the mucky haze in the air. Ozone injures tissue membranes, interferes with immune functions, and alters lung functions.[24] In addition, the EPA has reported that "tens of thousands of people die each year from breathing tiny particles (PM) in the environment."[25] Long-term studies in California have shown that particle pollution from diesel exhaust and soot may significantly reduce lung function in children.[26] In May of 2004, the EPA started an Asthma Awareness Month to educate the 20 million adults and children who suffer from asthma.[27]

The health danger from carbon monoxide occurs once it enters the bloodstream, where it interferes with a body's ability to deliver oxygen. Symptoms include reduced work capacity, poor learning ability, and visual impairment.[28] Sulfur dioxide is highly soluble, so it is rapidly absorbed in the upper respiratory system, where its major effect is to change the way the system functions. Because nitrogen oxides are not as soluble, they are absorbed into the bloodstream and can cause tissue damage in other areas besides the lungs.[29]

Nitrogen oxides have a major role in the formation of ozone and smog, which is responsible for millions of asthma attacks each year. Exposure to low levels of NO_2 over several hours may produce respiratory problems. Long-term exposure may cause permanent damage to the lungs.[30] Lead builds up in human tissues, blood, and bones. It can cause kidney, liver, and nervous-system disorders. In addition, lead can cause learning disabilities.[31]

Air toxins, such as benzene and mercury, may cause reproductive, developmental, and respiratory difficulties or illnesses. The EPA has listed benzene as a known **carcinogen** that can also cause various blood disorders and a depressed immune system.[32] Mercury is of such concern that government advisories have been issued to pregnant or nursing women and parents of young children, warning them of the hazards of eating certain types of fish, such as tuna. Mercury can cause damage to the brain and other organs and the immune system (Figure 2.3). One in six women of

Figure 2.3 In this 1973 photograph, a woman holds a victim of Minamata disease, which is a neurological syndrome caused by severe mercury poisoning. Minamata disease was first discovered in Minamata, Japan, after the Chisso Corporation dumped large volumes of wastewater containing methyl mercury into the ocean. More than 1,700 people have died from this chemical poisoning.

childbearing age in this country is estimated to have unsafe levels of mercury in her blood, putting about 600,000 newborns at risk because their nervous systems are sensitive to mercury.[33]

Visibility

Summer city air can prove that air pollution reduces visibility in many parts of the United States. However, because air and the pollution it carries travel, the problem has expanded to an increasing number of our national parks and refuges. The Department of the Interior has stated that reduced visibility "is the most ubiquitous air pollution—related problem in our national parks and refuges...all areas monitored for visibility show frequent regional haze impairment."[34] In addition, the Department of Interior estimates that more than 100 parks have unhealthy levels of ozone pollution. As one example, the EPA designated Sequoia-Kings National Park as having levels of ozone pollutions that threaten human health; between 1999 and 2003, the Park experienced 370 "unhealthful air days."[35]

Environmental

Global warming

A major **greenhouse gas** is carbon dioxide, a gas emitted by power and industrial plants in great quantities. Nitrous oxides, chlorofluorocarbons (CFCs), and other halocarbons are also greenhouse gases emitted by burning **fossil fuels**. These gases combined have been estimated to create as much global warming as the carbon-dioxide emissions.[36]

Hole in ozone

Ozone (O_3) is three atoms of oxygen linked together. If it forms close to the ground, it contributes to air pollution and health problems. However, the ozone that exists in a layer about 10 to 30 miles above Earth protects the planet from the Sun's harmful ultraviolet (UV-B) light. Ozone is created in a photochemical reaction. It is destroyed by chemicals used in air-conditioning, refrigeration, and some industrial processes. Chlorofluorocarbons (CFCs) and halons are examples of these chemicals.

In 1980, a massive hole was discovered in the atmosphere over Antarctica. Thinning of the ozone layer was also observed in other areas of the world (Figure 2.4). Depleted levels of ozone mean that more UV-B radiation reaches the ground. Exposure to this radiation increases the incidence of cancers, suppresses the immune system, and can cause cataracts. Damage to various plant and animal species has also been documented.[37]

Acid rain

The phrase "acid rain" does not have a good ring to it because acidic environments can kill living organisms. It is a problem throughout

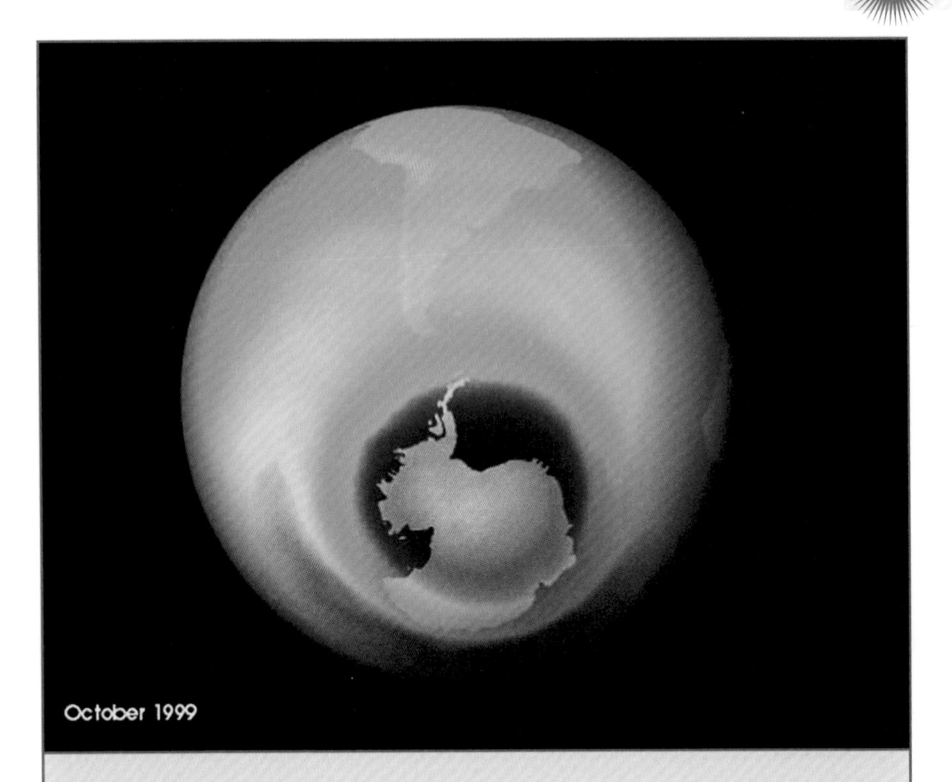

Figure 2.4 In this graphic from the Total Ozone Mapping Spectrometer (TOMS) Earth Probe, the areas of the depleted ozone layer *(blue)* over the Antarctic are clearly evident. In 2006, NASA and National Oceanic and Atmospheric Administration (NOAA) scientists reported that the ozone hole above polar regions in the Southern hemisphere is the largest to date.

the industrialized world and, indeed, for anyone that breathes the affected air. Acid rain was first given a name in 1872 in a book called *Air and Rain: The Beginning of a Chemical Climatology* by Robert Smith, who studied the air around coal-fired plants. About 50 years ago, it was identified in Europe, and in the 1980s, Canada and the United States began working together to reduce sources of acid-rain pollution.[38]

Acid rain is really more than rain. It includes both wet and dry deposits falling from the atmosphere. Rain and snow can carry acid rain, and acidic gases and particles can also fall to the ground.

Figure 2.5 Sulfur dioxide and nitrogen oxides emitted in the atmosphere lead to the onset of acid rain. In this photograph, the effect of acid rain is evident in the forests of Mount Mitchell in western North Carolina.

These deposits can be hundreds or even thousands of miles from the pollution source. Wet deposits seep into the soil or flow into surface waters. Dry deposits can be blown onto buildings or trees, later to be washed away by rain.

We know that acid rain is a result of sulfur dioxide and nitrogen oxides in the atmosphere (Figure 2.5). These gases react with the sunlight, water, and other chemicals to form sulfuric and nitric acids. Acid rain from pollutants generated in midwestern power plants falls in New York and New England, creating acidic lakes and soils. As an example, one-fourth of the surveyed Adirondack lakes in New York were too acidic to support fish, and 41% of all the surveyed lakes were acidic some or most of the time.[39] In the Green Mountains of Vermont, about half of the red spruce and beech trees have died in the past 40 years. Germany has seen more than half of the trees in the Black Forest die.[40]

Acid rain also damages buildings and monuments because the acids dissolve the stone. It changes water and soil chemistry, which affects the availability of plant and animal nutrients. The effects may be direct. Acid rain leaches the calcium out of red spruce needles, leaving the tree susceptible to freezing. Another direct effect is making lakes too acidic to support fish life. The effects may also be indirect. Changed environmental conditions can kill the food or the trees that fish, birds, or other animals eat or use for shelter.

Land and water

Toxic air pollutants can fall onto soils and surface waters. They can then be absorbed by both plants and animals, becoming part of the food chain. A recent news article noted the dangers of our eating fish with high concentrations of mercury. Fish-eating birds and mammals are also susceptible to the same harmful effects from these toxins. Nitrogen-oxide emissions from power plants and motor vehicles also contribute to gas, particle, and aqueous forms of nitrogen falling back to the ground, which causes acidification of lakes and soils. Adding nitrogen to coastal waters causes excessive algal growth. In addition, high mountain ecosystems, which are already sensitive to environmental changes, can be damaged by additional forms of nitrogen. Nitrous oxides also contribute to haze (ozone and smog), which damages crops and natural vegetation.[41]

HISTORY OF THE ISSUE

Is Air Pollution a New Problem?

Humans have been polluting the air ever since they started making wood fires. However, a few hundred years ago, humans began burning a new material—coal. Coal is a fossil fuel, made from ancient plants containing carbon, mercury, and sulfur. Burning coal releases these elements back into the atmosphere. Consider that 99% of all the humans who have ever lived on the planet are alive now. Even without doing a lot of math, one can see that this means a lot more air pollution. The Industrial Revolution gave us a new form of pollution—black smog, a mixture of soot and smoke. Although coal-fired industries still contribute heavily to air pollution, transportation vehicles—cars, trucks, ships, and planes—have created another new type of pollution. This type of city smog is the result of a photochemical reaction that creates ozone. It is the smog that creates the mucky brown haze that blankets cities in the summer.[42]

What Is the Past Policy?

Clean Air Act

Concern over air quality is not new. In 1955, the government allocated some funding to do research on air pollution. The Clean Air Act of 1963 designated more money for research and suggested that the states pass clean-air regulations. In 1970, public concern over the degrading environment provided impetus for major revisions to the 1963 law. The 1970 Clean Air Act (CAA) was a more stringent set of regulations and standards, providing the states with direction and assistance from the EPA to develop and enforce the new regulations. The intention was to regulate emissions to keep air pollution below harmful levels.

In 1977, Congress strengthened the Clear Air Act. A key feature of the amendments was setting a national goal to protect the scenic areas in our national parks and wilderness areas:

> ... the prevention of any future, and the remedying of any existing, impairment of visibility in protected national parks and wilderness areas which impairments results from manmade air pollution.[43]

Areas such as Grand Canyon, Yellowstone, Canyonlands, Mesa Verde, Rocky Mountain, and Great Sand Dunes national parks were designated for heightened protection.

In the 1990 amendments to the CAA, Congress continued to designate the states as responsible for bringing areas into compliance. However, the revisions also raised automobile emission standards and set new deadlines for reductions. The act required the use of best available control technology (BACT) to meet these

goals. Also included in this law was the reduction of the use of chlorofluorocarbons (CFCs), chemicals that destroy the ozone layer. The 1990 amendments identified 188 toxic air pollutants for the EPA to monitor and regulate. It gave the states more direction in implementing the regulations.[44]

Another 1990 amendment included Title IV, which was the first law in the United States to address the acid-rain issue. The law mandates specific reductions in nitrogen-oxide and sulfur-dioxide levels. In a creative approach to regulation, the EPA introduced the idea of cap and trade allowances. Each power plant is assigned emission allowances. Plants can then reduce their emissions and buy or trade their allowances from other industries.

In 1990, the EPA also adopted its Clean Air Act program of National Ambient Air Quality Standards (NAAQS). It identified and set health standards for the six "criteria pollutants"—sulfur dioxide, carbon monoxide, nitrogen oxide, lead, and ozone. The purpose was to provide the public with air-quality information through an air-quality index for different parts of the country. It was updated in 2003 to include data on fine particulate in the air. Information is available with annual reports of air quality in your neighborhood at www.epa.gov/air/data.

In 1997, EPA set new NAAQS for a form of air pollution known as "fine particles," or $PM_{2.5}$—particulate matter less than 2.5 microns in diameter.

New Source Review

The Clean Air Act mandates that all industrial and power plants built after 1970 use the best available technology to meet clean-air standards. However, many older plants are still operating. They have used a loophole in the regulations to avoid upgrading, stating that facility upgrades are simply maintenance and not expansion. Many power companies were sued and have had to pay substantial penalties [45] In 2003, the EPA changed the regulations in a way that would allow older facilities to avoid a new source review by upgrading in increments under 20% of plant size. Fourteen states and several environmental groups sued the EPA, which forced the EPA to review the regulations.

CURRENT ISSUES AND FUTURE CONSIDERATIONS

Although several major issues concerning air pollution are still to be resolved, some notable success has been achieved. Since 1970, monitored quantities of many air pollutants have decreased. New cars today emit about 75% less pollution than cars made before 1970—but the number of cars on the road has increased almost 50%.[46] Because Americans love to drive, cars and trucks remain the leading sources of air pollution in many urban areas.[47] Although improvements have been made in air quality, the continued poor quality in many cities has spurred another round of Clean Air Act amendments.

Another aspect to air-pollution control is being a responsible world citizen. By EPA estimates, about one-fourth of emissions from U.S. coal-burning plants are deposited within U.S. borders. The rest of U.S. emissions, 75%, "enter the global cycle," meaning winds carry U.S. pollution to other parts of the world.[48] The United States efforts to limit or not limit pollution have impacts around the world.

What Are the Economic and Environmental Concerns?

Clean air comes down to two issues: economics and the environment. One must consider how much controlling emissions and complying with regulations cost. However, the other cost consideration must assess the damage from air pollution to ecosystems and to human health. An EPA document outlining the 2005 Clear Air Interstate Rule (CAIR) states that the stricter emission regulations offer "dramatic health benefits at more than 25 times greater than the cost by 2015." It also states that by the year 2015, the CAIR will prevent 17,000 premature deaths, millions of lost school and work days, and tens of thousands of nonfatal heart attacks.[49] In addition, the report projects that as power-plant emissions decrease through 2020, the gross domestic product (GDP) steadily increases.

Any economic consideration should include a 2002 report issued by the United States Office of Management and Budget. The report states that over the 10-year period from 1992 to 2002, the four EPA rules that limit particulate matter and NOx emissions from heavy-duty engines and light-duty vehicles, and the Acid Rain rule, have had associated costs of between $8 billion and $8.8 billion per year. However, the benefits have been estimated to be $101 billion to $119 billion per year. Interestingly, the report cautioned that the costs actually could be a factor of 10 or more higher.[50] Some current estimates state that the health benefits from diesel-emission reductions outweigh the costs by 13 to 1.[51] In 2004, a $10 million investment by the West Coast Collaborative yielded more than $100 million in health benefits.

Environmental costs may be more difficult to quantify, but they are just as real. The sugar maple forests of Vermont are estimated to generate $110 million a year. These forests are showing decline from acid rain. Billions of dollars are generated in outdoor tourism and related outdoor recreational activities every year. To be successful, outdoor recreation—hunting, fishing, boating, and hiking are only a few examples—depends on healthy ecosystems, which means clean rivers, thriving forests, and clean air.

A key issue for the future involves cutting emissions from transportation. This goal can be accomplished by a variety of ways: cleaner fuels, more efficient engines, and energy conservation. The automotive industry is working on new technologies, including electric, hybrid, and **fuel-cell** vehicles (Figure 2.6). Equally important is reduction of emissions from coal-burning power and industrial plants. This reduction can be achieved by implementation of more stringent emissions standards on all plants, old and new. These standards must go beyond the current Clean Air Interstate Rule for the 28 eastern states to include the many new and aging plants in the western states.

Figure 2.6 On October 12, 2006, Steve St. Angelo, president of Toyota Motor Manufacturing, introduced the new Toyota Camry Hybrid. It is the first Toyota hybrid car to be introduced in North America. Hybrid vehicles cause less pollution and use less fuel.

What Are Current Policy Issues?

Clean Air Act

The EPA began revisions to the 1997 Clear Air Act several years ago. These revisions included a suite of clean-air standards aimed at reducing emissions from major polluters. This suite applies the following major rules: Clean Air Interstate Rule (CAIR), Clean Air Mercury Rule, Clean Air Nonroad Diesel Rule, Clean Air Ozone Rules, and Clean Air Fine Particle Rules. The first three rules directly address the air transport of pollution across state and national borders.

Although some policy makers feel that tighter standards are not necessary and inhibit economic growth, other groups feel cleaner air provides significant economic, health, and environmental benefits. Toward this end, several groups, including the American

Lung Association, Environmental Defense, and Earthjustice sued the EPA to include stronger standards in their plan released in March of 2005.[52]

Clean Air Interstate Rule

The Clean Air Interstate Rule (CAIR) was finalized on March 10, 2005. It applies to 28 states in the eastern United States and the District of Columbia. The CAIR addresses the problem of pollutant transport by air and provides a strong regulatory tool for use of the Clean Air Act to protect human health and the environment. When implemented, CAIR will see the largest reduction in air pollution in the previous decade. By 2015, a permanent cap on SO_2 and NOx emissions in the eastern United States will be in place.[53] The EPA projects that this cap will result in more than $100 billion in health and visibility benefits per year.

Clean Air Nonroad Diesel Rule

In 2001, the EPA strengthened its national standards for truck and bus diesel engines. In May of 2004, the EPA implemented the Clean Air Nonroad Diesel Rule, with the intention of cutting emissions from nonroad sources by 90%, to be in full effect by 2015. As then EPA administrator Mike Leavitt stated, "We're able to accomplish this in large part because of masterful collaboration with engine and equipment manufacturers, the oil industry, state officials, and the public health and environmental communities."[54] On July 7, 2005, an amendment to these regulations was signed that established low-sulfur requirements for fuel, beginning in 2006.[55]

Clean Air Mercury Rule

In 2003, the EPA proposed changes to the Clean Air Act that would allow power plants to avoid mercury-emissions standards. Eleven states and several environmental groups opposed the changes, with the Connecticut Attorney General, Richard Blumenthal, stating that "This action makes a mockery of environmental justice and the EPA's mandate to protect public health. The EPA's attempt to reverse its own mercury emission rules underscores how the power industry has hijacked the agency."[56]

In March of 2005, the EPA released its new rules that would allow coal and oil-direct utility units to actually slow their clean-up efforts for more than a decade. The new rule was legally challenged in federal court. In addition, on July 18, 2005, 32 U.S. senators filed a "Mercury Rule Discharge Petition," requesting congressional disapproval of this new EPA mercury rule.[57] On September 13, 2005, the Senate narrowly rejected this resolution (47 to 51) to stop increased mercury pollution.

Clean Air Ozone Rule

On June 15, 2004, the EPA issued the final rule to implement the Eight-Hour Ozone National Ambient Air Quality Standard—Phase I. The rule establishes the classification scheme for areas that do not meet air-quality standards, called "nonattainment" areas. The rule also mandates that states continue to follow the regulations with respect to existing one-hour ozone requirements. Deadlines for meeting these health-based, eight-hour ozone standards (as determined by the fourth highest eight-hour daily maximum at any single monitoring site, averaged over a three-year period) are from 2007 to 2021, depending on the area. Some states have chosen to meet the deadlines ahead of schedule.

Business, State, and Local Initiatives

Not only are many businesses, large and small, initiating efforts to reduce emissions of global-warming gases, they are also involved in programs to improve air quality. One example is Federal Express. In April of 2005, FedEx announced plans to use up to 75 hybrid diesel-electric trucks in the next year. Hybrids emit 90% less pollutants and increase fuel efficiency by 50%. The company, which has been working with Environmental Defense since 2000, currently has 18 hybrid trucks in use in Sacramento, New York City, Tampa, and Washington, DC.[58]

An example of an ambitious partnership among federal, state, and local governments; businesses; and environmental groups from California, Oregon, Idaho, Washington, Alaska, British Columbia, and Mexico is the West Coast Diesel Reduction Collaborative. Its purpose is to reduce diesel emissions. Business partners include more than 130 companies, among them Toyota, Union Pacific Railroad, and General Motors. In 2004, the collaborative gave $9 million to meet a $1 million EPA grant to reduce diesel emissions. This project resulted in more than $100 million in health benefits.[59] Automakers are responding to changes in current regulations and looking ahead to new ones by offering more than 100 models of vehicles with EPA-estimated ratings of 30 miles (48.2 km) per gallon or more.[60]

States are also involved in ambitious clean-air initiatives. North Carolina has led the way in reducing air pollution from coal-fired plants by enacting the 2002 Clean Smokestacks Act. The act mandates that plants must reduce NOx emissions by 77% by 2009 and reduce SO_2 emissions by 73% by 2013.[61] In 2003, Illinois kicked off its "Illinois Clean School Bus Program" as part of Children's Health Month. The program was designed to significantly reduce emissions from diesel-powered school buses.[62] In addition, the Illinois EPA sponsored a gas can exchange and lawn mower buyback events during 2005 to help reduce VOC emissions. California, always concerned about its growing population and huge number of automobiles, was the first to pass auto-emissions standards in 1966. In 1990, the state approved standards for cleaner-burning fuels and low-emissions and zero-emissions vehicles. Recently, California passed more stringent auto-emissions standards, but the state was sued by automobile makers. The Bush administration agreed that the state was acting out of its legal realm. The headlines in California are important to watch, as the battle over states' rights continues to be fought in the clean-air arena.

In 2005, the EPA announced a $7.5 million grant program to help school districts reduce emissions from school buses that have diesel engines.[63] Wisconsin responded to the 1996 EPA news that six counties had one of the most severe ground-level ozone problems in the country by forming a coalition of more than 260 businesses, schools, government, and community agencies working together to improve air quality. The Wisconsin Partners for Clean Air (WPCA) works to help implement the state plan for the Clean Air Act.[64]

In addition, cities are working toward cleaner air. During the summer of 2005, the city of Providence, Rhode Island, offered free bus service on "ozone alert days" to decrease the number of automobiles on the roads. New York City launched an impressive initiative in 2000. Other public transportation agencies have taken steps toward improving air quality by using alternatives to standard diesel fuel in buses. The Chicago Transit Authority and the Massachusetts Bay Transportation Authority have been powering buses on ultralow-sulfur diesel fuel for several years. Cincinnati Metro and Central Oklahoma Transportation and Parking Authority have been using biodiesel fuels (domestically produced, renewable fuel made from non–fossil fuel sources) in buses and trolleys, respectively, for about five years.

WHAT THE INDIVIDUAL CAN DO

1. Save energy in the home.
   - Recycle. This activity can save 70 to 90% of energy required to make new products.
   - Plant trees. They insulate a home and clean the air.
   - Turn computers and monitor off when not in use.
   - Cool and heat less. Even a two-degree difference helps.
   - Conserve water.
   - Replace old appliances with energy-efficient ones. (Buy EnergyStar rated).
   - Insulate your home. Use storm windows.
   - Clean your furnace and change the filters regularly.
   - Use compact fluorescent lighting, which can use up to 75% less energy.
2. Drive smart.
   - Drive a fuel-efficient vehicle.
   - Use efficient tires and keep the recommended air pressure in them.
   - Consolidate driving trips.
   - Consider using mass transit.

REFERENCES

1. "State of the air report," American Lung Association. Available online. URL: http://lungaction.org/reports/stateoftheair2005.html. 2005.
2. "Stop blowing smoke in the heartland," Environmental Defense. Available online. URL: http://www.environmentaldefense.org/documents/3887_blowingsmoke.pdf. 2004.
3. "Mercury: Basic information," Environmental Protection Agency Factsheet. Available online. URL: http://www.epa.gov/mercury/about.htm.
4. Godish, Thad. *Air Quality*. 4th Edition. Boca Raton, Fla.: Lewis Publishers, 2004.
5. Godish, *Air Quality*, p. 9.
6. Turco, Richard P. *Earth Under Siege: From Air Pollution to Global Change*. New York: Oxford University Press. 2002.
7. Godish, *Air Quality*, p. 25.
8. Wright, Richard T. *Environmental Science*. 9th Edition. Upper Saddle River, N.J.: Prentice Hall, 2005.
9. "The ambient air monitoring program," The Environmental Protection Agency. National Ambient Air Quality Standards (NAAQS). Available online. URL: http://www.epa.gov/air/criteria.html. 2005.
10. Godish, *Air Quality*, p. 39.
11. "Nitrogen dioxide," Environmental Protection Agency Factsheet. Available online. URL: http://www.epa.gov/airtrends/nitrogen2.html. 2005.
12. "Coal-fired power plants are big contributors to sooty particle pollution in eastern states," Environmental Defense. Available online. URL: http://www.environmentaldefense.org/article/cfm?ContentID=3842). July, 2004.
13. Audesirk, Teresa, and Gerald Audesirk. *Biology: Life on Earth*. 4th Edition. Upper Saddle River, N.J.: Prentice Hall, 1998.
14. Decker, Hilary, et al. *Closing the Diesel Divide*. New York: Environmental Defense and the American Lung Association, 2003.
15. Ryan, John C. "Dust in the wind: Fallout from Africa may be killing coral reefs an ocean away. World Watch Institute. Available

online. URL: http://www.worldwatch.org/pubs/mag/2002/151/. January/February 2002.

16. "More details on carbon monoxide: National air quality and emissions trends report." EPA Air Trends. Available online. URL: http://www.epa.gov/airtrends/carbon2.html.
17. Turco, *Earth Under Siege*, p. 154.
18. Wright, *Environmental Science*, p. 581.
19. "More details on lead: National air quality and emissions trends report. Toxic air pollutants," EPA Air Trends. Available online. URL: http://www.epa.gov/airtrends/lead2.html. 2005.
20. "About air toxins," Technology Transfer Network Air Toxic Web site. EPA. Available online. URL: http://www.epa.gov/ttn/atw/allabout.html. 2005.
21. Godish, *Air Quality*, p. 156.
22. Ibid.
23. "Air pollution, heart disease, and stroke," American Heart Association. Available online. URL: http://www.americanheart.org/presenter.jhtml?identifier=4419. 2004
24. Godish, *Air Quality*, p. 169.
25. Godish, *Air Quality*, p. 156.
26. "Recent research findings: Health effects of particulate matter and ozone air pollution," California EPA Air Resources Board and American Lung Association of California. Available online. URL: www.arb.ca.gov/research/health/fs/pm-03fs.pdf. January 2004.
27. "May is allergy awareness month," Environmental Protection Agency. Available online. URL: http://www.epa.gov/newsroom/allergy_month.htm. May, 2004.
28. "More details on carbon monoxide: National air quality and emissions trends report," EPA Air Trends. Available online. URL: http://www.epa.gov/airtrends/carbon2.html.
29. Godish, *Air Quality*, pp. 158, 167.
30. "Nitrogen oxides," Environmental Protection Agency. Available online. URL: www.epa.qld.gov.au/environmental_management/air/air_quality_monitoring/air_pollutants/nitrogen_oxides/. 2005.
31. "More details on lead: National air quality and emissions trends report. Toxic air pollutants," EPA Air Trends. Environmental

Protection Agency. Available online. URL: http://www.epa.gov/airtrends/lead2.html. 2005.
32. Wright, Richard T. *Environmental Science*. 9th Edition. Upper Saddle River, N.J.: Prentice Hall, 2005.
33. "Bush mercury policy threatens the health of women and health,". Natural Resources Defense Council Backgrounder. Available online. URL: http://www.environmentaldefense.org/system/templates/page/subissue/cfm?subissue=20.
34. "Code red: America's five most polluted national parks," National Parks Conservation Association. Available online. URL: http://www.npca.org/across_the_nation/visitor_experience/clear_air/code_red/default.asp.
35. Ibid.
36. Wright, *Environmental Science*, p. 551.
37. "Stratospheric ozone," Environmental Protection Agency. Available online. URL: http://www.epa.gov/airtrends/strat.html. 2005.
38. Turco, *Earth Under Siege*, p. 260.
39. Audesirk, *Biology*, p. 908.
40. "Clearing the haze from western skies," Environmental Defense. Available online. URL: www. Environmentaldefense.org/go/westernhaze. June 2005.
41. Turco, *Earth Under Siege*, p. 272.
42. "Clearing the haze from western skies," Environmental Defense. Available online. URL: www. Environmentaldefense.org/go/westernhaze. June 2005.
43. Clean Air Act 2005. Environmental Protection Agency. Available online. URL: http://www.epa.gov/oar/caa/contents.html.
44. Wright, *Environmental Science*, p.599.
45. Ibid., p. 595.
46. "Transportation, sprawl, and health," Environmental Defense Factsheet. Available online. URL: www. ED.org/go/transportation. 2004.
47. "Mercury: Basic information," Environmental Protection Agency Factsheet. Available online. URL: http://www.epa.gov/mercury/about.htm.

48. CAIR 2005. Environmental Protection Agency. Available online. URL: http//:www. Epa.gov/cair/basic/html. 2005
49. Ibid.
50. "Informing regulatory decisions: 2003 report to Congress on the costs and benefits of federal regulations," Available online. URL: http://www.whitehouse.gov/omb/inforeg/2003_cost-ben_final_rpt.pdf.
51. "West Coast collaborate: Public-private partnership to reduce diesel emissions," West Coast Diesel Reduction Collaborative Factsheet. Available online. URL: http://www.westcoastdiesel.org/. 2005.
52. News Release. "EPA clean air plan must ensure no child left behind," Environmental Defense, April 15, 2005.
53. EPA. "Clean interstate rule," Available online. URL: http://www.epa.gov/cair/basic/html. 2005.
54. "EPA announces landmark clean air interstate rule," Available online. URL: http://www.epa.gov/boston/pr/2005/mar/sr050306.html. 2005.
55. EPA Factsheet. "Diesel fuel," Available online: URL: http://www.epa.gov/otaq/regs/fuels/diesel/diesel.htm. 2005.
56. Press Releases. Office of NY State Attorney General Eliot Spitzer. Available online. URL: http://www.oag.state.ny.us/press/2004/jun/jun28b_04.html. June 28, 2004.
57. Press Release. "Mercury rule discharge petition filed," U.S. Senator Patrick Leahy. Available online. URL: http://leahy.senate.gov/press/200507/071805.html.
58. Press Release. "FedEx announces plans to add up to 75 hybrid trucks to fleet," Available online. URL: http://www.environmental-defense.org/partnership_project.cfm?projectID=3.
59. West Coast Diesel Emissions Reduction Collaborative. Available online. URL: http://www.epa.gov/Region9/air/westcoastdiesel/feature.html. June 2005.
60. Auto Alliance. "Fuel efficiency," Available online. URL: http://www.autoalliance.org/fuel/fuel_efficiency.ph. 2005.
61. North Carolina Current Legislation. Available online. URL: http://daq.state.nc.us/news/leg/. 2005.

62. Press Release. "Governor announces kick-off of his statewide clean school bus program," Available online. URL: www.epa.state.il.us/air/cleanbus/. 2003.
63. EPA. "National clean diesel campaign," Available online. URL: http://www.epa.gov/diesel/. 2005.
64. "What is WPCA?" Wisconsin Partners for Clean Air. Available online. URL: http://www.cleanairwisconsin.org/aboutus.php. 2005.

FURTHER READING

Books

Godish, Thad. *Air Quality.* Boca Raton, Fla.: Lewis Publishers, 2004.

Turco, Richard P. *Earth Under Siege: From Air Pollution to Global Change.* New York: Oxford University Press, 2002.

Web Sites

American Heart Association

http://www.americanheart.org

American Lung Association

http://www.lungusa.org

Environmental Protection Agency

http://www.epa.gov

Natural Resources Defense Council

http://www.nrdc.org/air/pollution/default.asp

Global Climate Change

WHAT'S IN THE NEWS?

The glaciers are melting, the sea level is rising, and Earth is getting warmer. These are observable facts. The warming and cooling cycles of our planet's climate have been happening for millions of years, as evidenced by studying such clues as fossil records and air pockets in ice-core samples. However, as in the case of species extinction, the newsworthy twist is the *rate* at which these changes are now happening. The rate of change cannot be explained by natural causes alone.[1] If the glaciers continue melting, sea level continues rising, and temperatures continue warming at the present rate, scientists predict serious economic and environmental problems across the globe.

"The greenhouse effect is the most significant economic, political, environmental and human problem facing the twenty-first century," said Timothy Wirth, former U.S. senator and undersecretary of state for global affairs.[2]

Many experts believe that climate change is being accelerated by human activities, with burning fossils fuels and deforestation as the main contributors to increased greenhouse, or warming, gases in the atmosphere. The United Nations Intergovernmental Panel on Climate Change (IPCC) has predicted that the average global temperature will increase between 2.5°F and 10.4°F over the next century.[3]

Why does climate change matter to you? Our health depends on clean supplies of water and air, productive agriculture, healthy fisheries, and productive oceans. Science is telling us that these things are being negatively affected by climate change and the potential exists for significant negative effects to our planet. The question remains how policy makers at all levels will respond.

THE SCIENCE

What Is Climate Change?

Climate change is a natural part of Earth's history. However, Earth is 4.6 billion years old, and natural climate changes have occurred slowly over tens of thousands, and even millions, of years. As we saw in the case of species extinction, the problem is not that it happens, but the rate at which it is happening now.

Historically, climate changes have been associated with changes in the levels of carbon dioxide in the atmosphere. The warmer periods had higher carbon-dioxide levels, whereas the cooler periods had lower carbon-dioxide concentrations.[4]

What Is in Earth's Atmosphere?

Earth is blanketed by a layered atmosphere (Figure 3.1). The lower layer, the troposphere, hugs the planet, rising about 6.8 miles (11 km). This layer keeps Earth warm enough to sustain life. The upper layer, the stratosphere, wraps itself around the troposphere and extends out to 31 miles (50 km) above Earth.

The atmosphere is primarily made of three gases: 78.9% nitrogen, 20.95% oxygen, and 0.93% argon. The remaining volume is a mere 0.1%, consisting of carbon dioxide, methane, nitrogen oxides, carbon monoxide, chlorofluorocarbons (CFCs), and ozone. These trace gases, along with water vapor, are commonly called greenhouse gases because they trap heat in the atmosphere.

Are Climate and Weather the Same?

The term "climate change" is defined by the United Nations Framework Convention on Climate Change to mean a change that is caused by human activity, that alters the global atmosphere, and that is observed over comparable time periods as natural climate

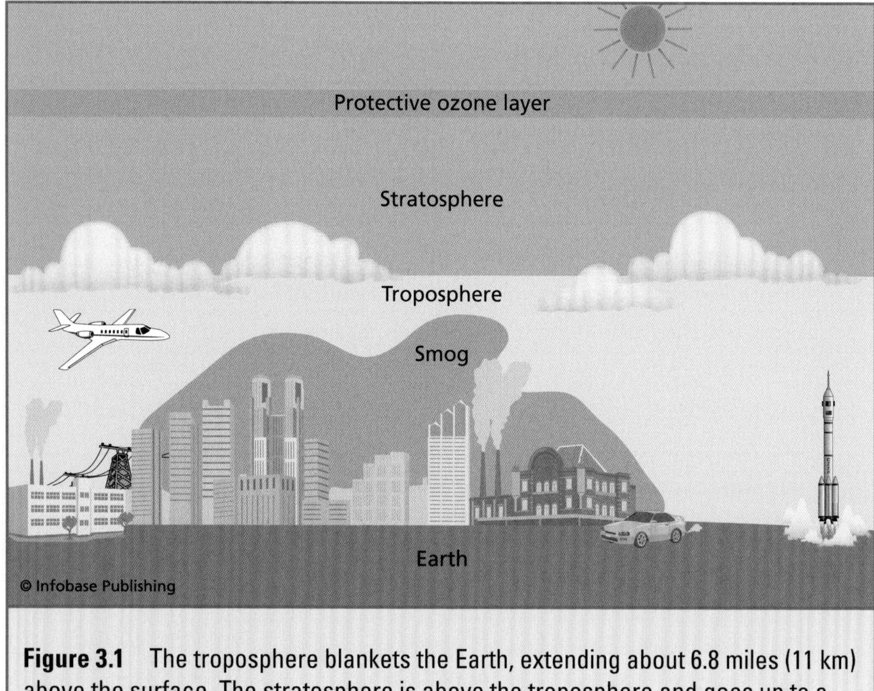

Figure 3.1 The troposphere blankets the Earth, extending about 6.8 miles (11 km) above the surface. The stratosphere is above the troposphere and goes up to a height of 31 miles (50 km).

changes.[5] Weather, on the other hand, is defined as the state of the atmosphere with respect to such factors as wind, temperature, moisture, and pressure at a specific point in time. Therefore, climate is considered the overall pattern of weather.

Climate and weather are both driven by solar energy or the way in which heat is distributed over Earth. The Sun's heat strikes Earth at different angles, or angles of incidence. More solar heat is delivered to the tropics around the equator because the rays strike at a more direct angle. Physics tells us that the warmer air rises and naturally moves toward the cooler air masses around the poles. These differences in air temperatures, together with the rotation of Earth, cause winds. Hot, moist air over the equator rises and then sinks as it cools over the poles. This transfer of energy is one force that drives the huge circulation patterns in both the oceans and the air. Huge ocean currents, such as the Gulf Stream in the

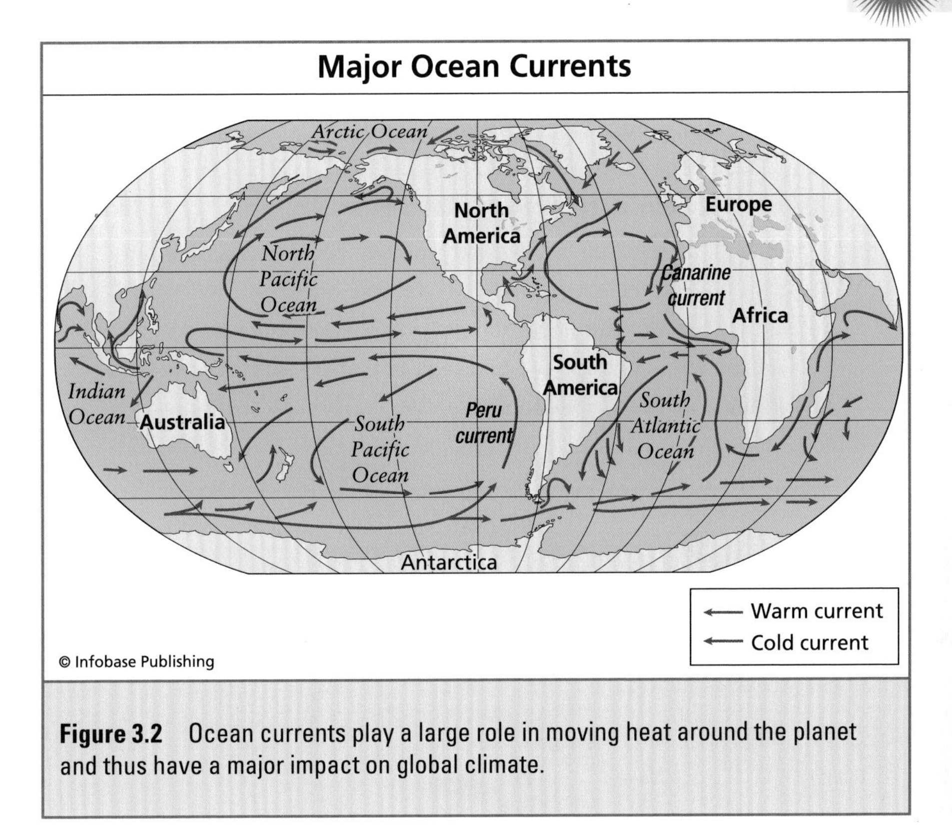

Figure 3.2 Ocean currents play a large role in moving heat around the planet and thus have a major impact on global climate.

Atlantic Ocean, also are driven by differences in water temperature and salinity (Figure 3.2). Significant changes in any of these factors will change the weather patterns.

What Is the Carbon Cycle?

To understand the issue of global climate change in the news today, one must understand the way carbon cycles on our planet. The question for scientists and policy makers today is how much carbon can be naturally cycled out of the atmosphere. The question is important because humans are adding carbon at an accelerated rate.

Carbon, like the elements sulfur, nitrogen, and oxygen, continuously moves around the planet in complex cycles, through living and nonliving materials (Figure 3.3). Carbon is found in various forms and in many places—from the air to fossil fuels to the DNA in genes. Carbon occurs as carbon dioxide (CO_2), carbonic acid (H_2CO_3), bicarbonate (HCO_3), carbonate (CO_3), and organic matter (CH_2).[6] Some processes release carbon into the atmosphere, and others use or store it, serving as **carbon sinks**.

Carbon is released as organic matter decomposes. Human activity, such as cutting down forests and burning fossil fuels, accelerates the amount of carbon dioxide released into the air.[7]

Carbon sinks include the oceans, the planet's largest reservoir of carbon. The oceans hold, or sequester, carbon in sediments, in

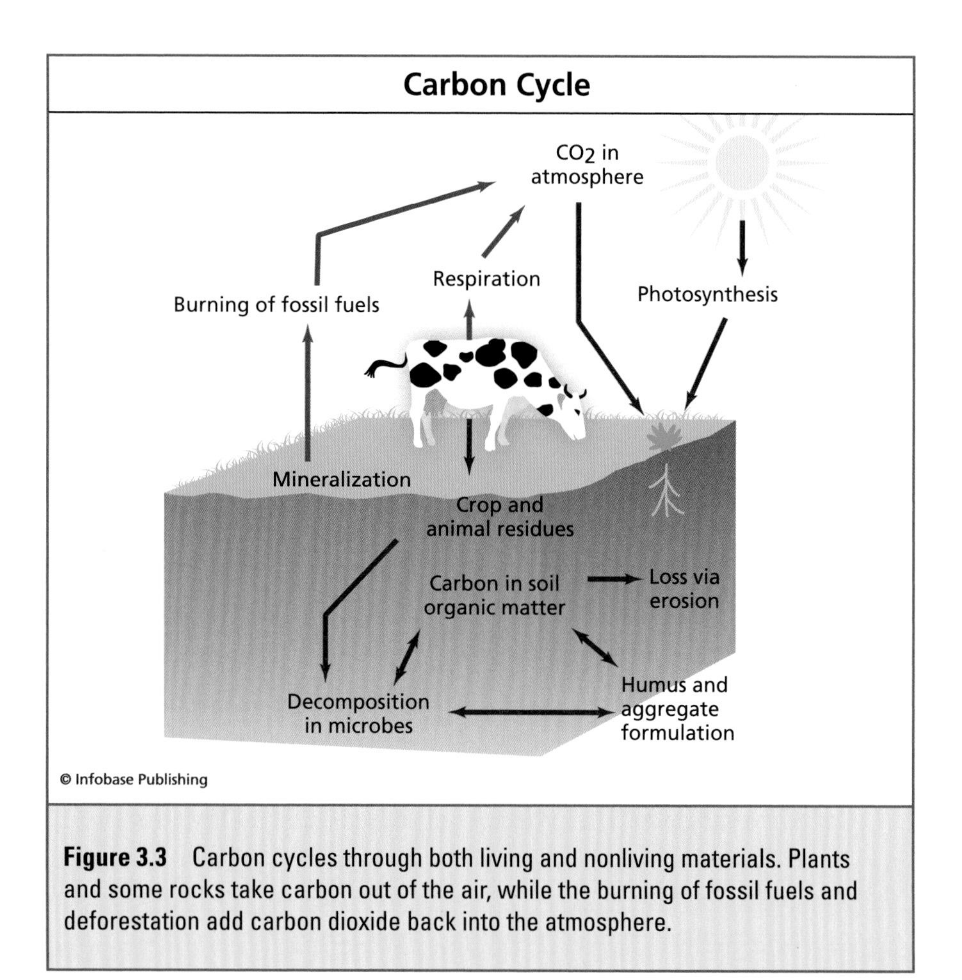

Figure 3.3 Carbon cycles through both living and nonliving materials. Plants and some rocks take carbon out of the air, while the burning of fossil fuels and deforestation add carbon dioxide back into the atmosphere.

organisms, and dissolved in the water. The land is the second-largest carbon reservoir, storing carbon in rocks and organic matter (plants and animals).

The carbon cycle is complex and not fully understood. For example, how climate change will affect the amount of carbon dioxide dissolved in the ocean or taken up by algae in the ocean is unclear. Another complexity of the carbon cycle is that it occurs at different rates. Some carbon has been stored for millions of years as fossilized organic carbon—gas, coal, and oil. Other carbon is captured from dissolved water as mollusks (clams) make their shells. It is later deposited as limestone, which may, over long periods of time, dissolve in water again. However, some carbon cycling occurs more quickly, as when plants use carbon dioxide in photosynthesis to make complex sugars, and leaves decompose the next season.

What Affects Climate Change?

The United Nations Framework Convention on Climate Change has defined climate change as a change of climate attributed directly or indirectly to human activity that alters the composition of the global atmosphere and is observed over comparable time periods.[8] Human factors that can cause variations in climate are fluctuations in concentrations of greenhouse gases and changes in land use.

The Greenhouse Effect

Our planet is not a ball of ice like Pluto. This is because the Sun emits huge quantities of radiation that penetrates our atmosphere. Most of this energy is ultraviolet light, or short waves. The waves are then reradiated back into the atmosphere, but this time as longer wavelengths. These longer wavelengths are absorbed by gases in our atmosphere. Just as the heat trapped by glass warms the inside of a car or a greenhouse, gases in our atmosphere absorb heat and create a blanket of warmer temperatures. This

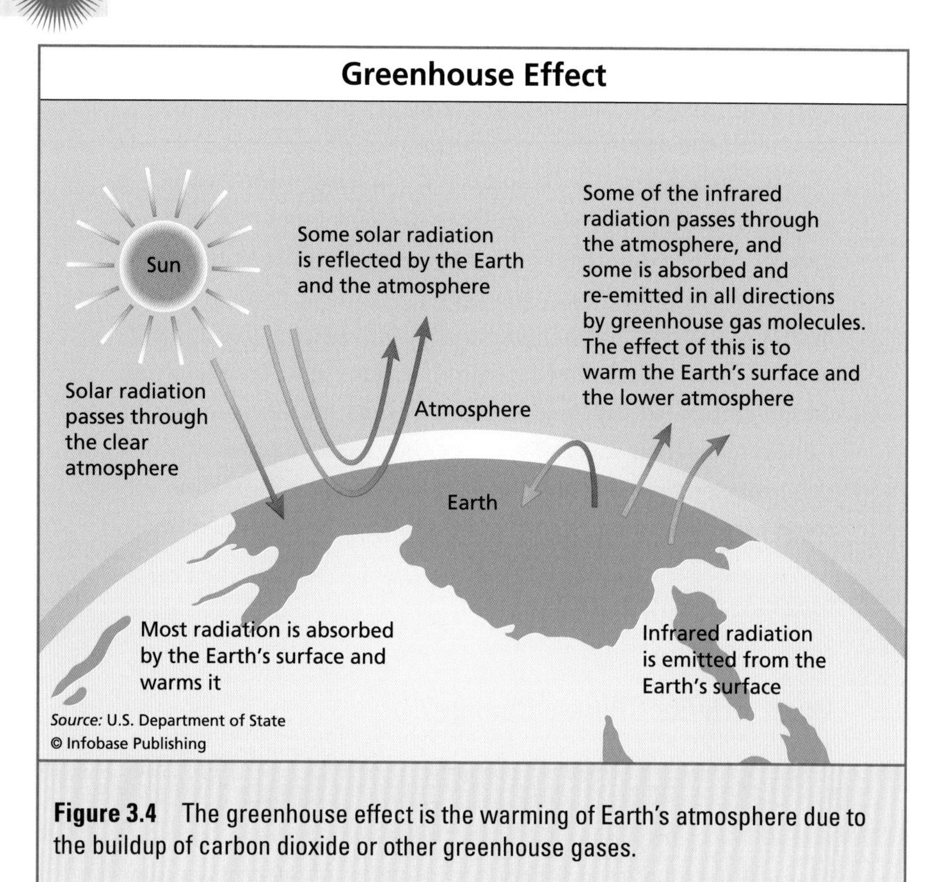

Figure 3.4 The greenhouse effect is the warming of Earth's atmosphere due to the buildup of carbon dioxide or other greenhouse gases.

phenomenon, called the greenhouse effect, created the conditions that allow life on our planet. Without these gases absorbing, or trapping, the heat, our planet would stay at approximately 0° Fahrenheit.[9] The greenhouse effect becomes a problem when increased concentrations of greenhouse gases absorb more heat, thus increasing global temperatures (Figure 3.4).

Greenhouse Gases

Five gases trap the most heat in the atmosphere: carbon dioxide, methane, chlorofluorocarbons (CFCs), nitrous oxide, and water vapor. These heat-trapping gases are called greenhouse gases

(GHGs). Each gas absorbs different amounts of heat. The IPCC developed the Global Warming Potential (GWP) concept to compare the ability of each greenhouse gas to trap heat in the atmosphere relative to other gases. For example, methane traps more than 21 times more heat per molecule than does carbon dioxide, whereas nitrous oxide absorbs 270 times more heat per molecule than does carbon dioxide.[10]

The atmospheric concentration of these gases can be directly affected by human activities. Since preindustrial times, atmospheric methane has increased 151 times.[11] The IPCC issued a report in 2001 with the following conclusion:

> In light of new evidence and taking into account the remained uncertainties, most of the observed warming over the last 50 years is likely to have been due to the increase in greenhouse gas concentrations.[12]

Greenhouse gases are measured in pounds entering the atmosphere. For example, the energy needed to power the appliances in a single-family house creates about 28,000 pounds of carbon dioxide a year. The power plants needed to supply all the homes in the United States generate 1,350,300,000 tons of carbon dioxide a year.[13] Each person in the United States generates approximately 6.6 tons (almost 15,000 pounds of carbon equivalent) of greenhouse gases each year. More than 80% of this carbon dioxide is from burning fossil fuels to power homes and transportation.[14]

Carbon dioxide is a significant factor in trapping heat in the atmosphere. Researchers began measuring the levels of different greenhouse gases in the late 1950s, taking their readings at a remote site on Hawaii's highest volcano. The data indicated an increase in the buildup of carbon-dioxide levels in our atmosphere.[15] Carbon-dioxide concentrations have risen 19% since the first Hawaii readings, and the concentrations have increased 35% since the beginning of the industrial age. Since 1960, the average annual rates of carbon-dioxide increases have doubled.[16]

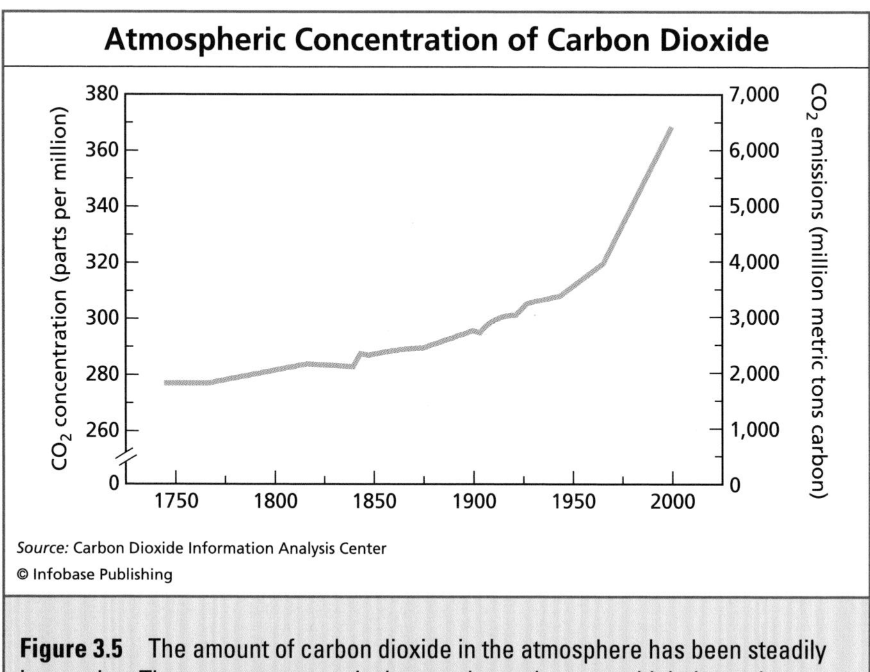

Figure 3.5 The amount of carbon dioxide in the atmosphere has been steadily increasing. The current concern is the accelerated rate at which the carbon dioxide concentration is rising and causing the Earth to warm.

Fossil Fuels

Fossil fuels (coal, oil, and gas) are huge reservoirs of carbon formed millions of years ago. Coal beds were formed when layers of plants were deposited in prehistoric swamps, becoming organic peat. Heat and pressure changed the layers of organic peat into coal. Gas and oil deposits formed as plants and animals accumulated on ancient sea bottoms. Because plants and animals are carbon based, burning fossil fuels releases large amounts of carbon dioxide into the atmosphere (Figures 3.5 and 3.6).

Since the beginning of the industrial age, the levels of carbon dioxide in the atmosphere have dramatically increased. Electricity generation from the burning of fossil fuels is the largest source of human-induced carbon-dioxide emissions, totaling about 37% of the global emissions.[17]

Changes in Land Use

Currently, up to 25% of the carbon-dioxide increases in the atmosphere can be attributed to land use changes, including forest cutting and agriculture.[18]

Deforestation

Cutting of forests, or deforestation, especially in the biologically productive tropical rain forests, takes away a major carbon sink because the trees and plants are no longer taking in carbon as they photosynthesize. The tropical rain forests are disappearing at a rate of 214,000 acres (86,000 hectares) a day, which is 78 million acres (31 million ha) per year.[19]

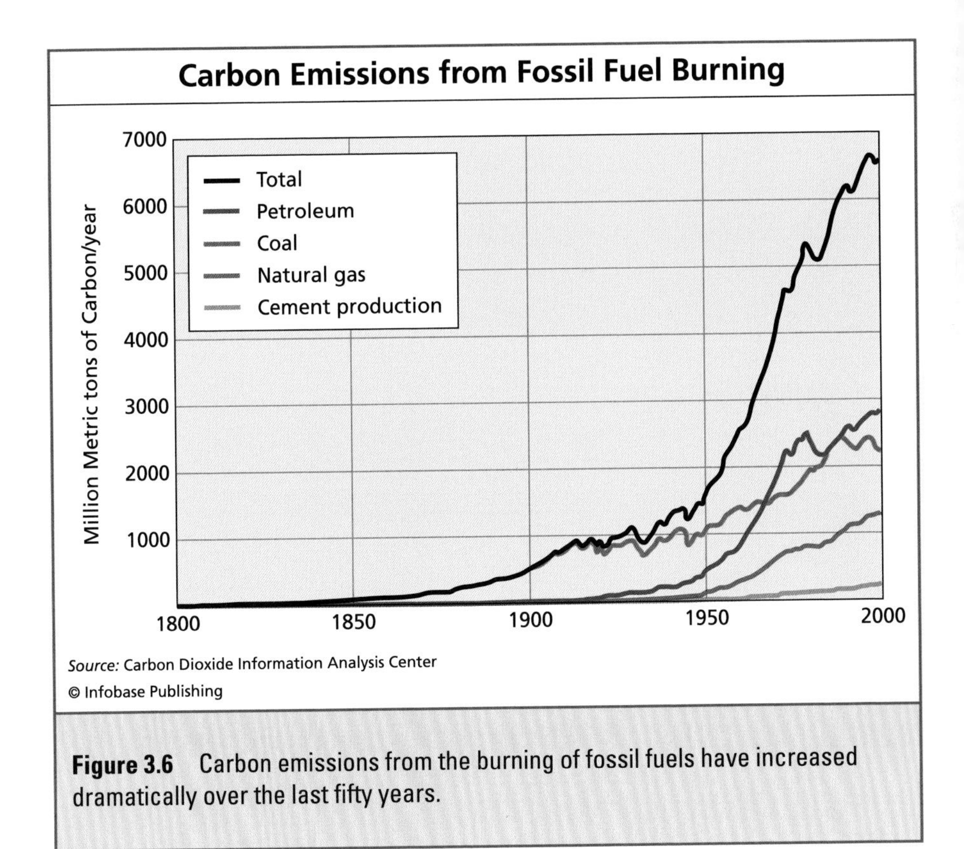

Figure 3.6 Carbon emissions from the burning of fossil fuels have increased dramatically over the last fifty years.

Agriculture

The second most significant greenhouse gas is methane, which is produced when organic matter decomposes without oxygen, as in rice paddies or landfills. Methane is also produced by cows and sheep (burping and flatulence). Methane is now more abundant in Earth's atmosphere than at any time in the past 400,000 years.[20] As the world's population continues to grow, the agricultural operations of cows, sheep, and rice paddies will continue to increase methane levels in the atmosphere.

Does Scientific Evidence of Climate Change Exist?

Temperature

The surface temperature of Earth has increased by about 1°F in the past century, according to the National Academy of Sciences.[21] Again, the rate of change is what has many scientists concerned; since 1976, the rate has tripled from what is was for the past century. In 2004, the average annual global temperature was the fourth warmest since 1880; the others occurred in 1998, 2002, and 2003.[22] Scientific evidence shows that the twentieth century was the warmest in the past 1,000 years. The 1990s were the warmest decade in the past millennium.[23]

Precipitation

The surface of the planet is more than two-thirds water, which means that increased temperatures will cause a significant increase in evaporation. Because water vapor is a greenhouse gas, increased evaporation increases the greenhouse effect. Water that evaporates eventually falls as rain or snow. Some changes in rainfall patterns have already been detected. Rain-gauge readings from more than 5,000 gauges on six continents show that average annual rainfall has increased almost an inch in the past 100 years.[24] The rainfall changes are not in any uniform pattern. Data for the Great Lakes region for the past 50 years show more frequent and more severe rainstorms.[25]

Another example is the change in precipitation from snow to rain during the winter months in the Arctic regions—an increase of 50% over the past 50 years in western Russia.[26] Snow provides water storage, which, unlike rain, slowly melts and releases over time. Many people and ecosystems depend on this slower release of water provided by mountain snow.

Sea Level

The sea level has been rising since the end of the last Ice Age, about 18,000 years ago. However, it is presently rising at a faster rate than over the past several thousand years.[27] During the past 100 years, the average sea level has risen by almost 8 inches (20.3 centimeters). On a local level, consider that Louisiana currently is losing about 2 acres of land every hour, or 34 square miles every year as the sea level rises and the land is submerged.[28] The coasts of Nova Scotia and Maine have seen a sea level rise of 1 to 2 feet (.3 to .6 m) in the past 250 years.[29]

Glaciers

Scientific evidence that the glaciers are melting is abundant. In 1850, more than 150 glaciers were in Glacier National Park. Today, about 35 glaciers remain, and scientists predict that these may all melt by 2030.[30] According to the National Park Service, since the early 1980s, many of the glaciers on Mount Rainier have been thinning and retreating.[31]

In addition, significant changes in the sea ice in the higher latitudes have been reported using recently declassified information from U.S. and Russian submarine data. Data indicate that sea ice in the central Arctic has thinned since 1970, and a 10 to 15% decrease in summer sea ice has occurred in the Arctic region.[32] According to scientists, the fastest moving glacier in the Arctic, the Sermeq Kujalleq Glacier, is melting and breaking up (Figure 3.7). In 2003, the glacier front was 6.8 miles (11 km) east of its usual position.[33] A new study outlined in the April 2005 issue of the journal *Science* found that 84% of the glaciers in a part of Antarctica have gotten smaller in the past 50 years.[34]

Figure 3.7 The Sermeq Kujalleq glacier *(above)*, located in Ilulissat, Greenland, is one of the most active glaciers in the world. In 2004, UNESCO placed the glacier on the World Heritage List, just months before it was reported to be receding in size.

Species and Ecosystems

Altered patterns of rainfall and increased temperatures are already having effects on some plant and animal habitats, as well as their reproductive cycles and ranges.[35–37] In addition, increased water temperatures are damaging coral reefs, the most biologically rich marine ecosystem on Earth. Warmer waters cause coral bleaching and a higher incidence of diseases in the reefs.[38] In southern New England, lobster catches have plummeted because of parasites and heat stresses of warming waters.[39]

Desertification

Desertification refers to the degrading of land in arid, semiarid, and dry subhumid areas, which cover about a third of the planet's

land area. Desertification is caused by human activities, such as clearing land, overgrazing land, or bad irrigation techniques. However, the second major cause of desertification is climate change. More than 250 million people are currently affected by desertification, with more than a billion more at risk.[40]

How Do Scientists Predict Climate Change?

People cannot control nature, yet with improved technology, scientists are getting better at predicting some natural events, such as earthquakes and volcanic eruptions. Scientists predict climate change by looking at the past, using data from the study of fossils, tree rings, and air trapped in ice cores, and know that some climatic fluctuations have occurred in the past. However, climate patterns tell us that Earth's climate is changing more rapidly now. How do we know what the climate will be like in a 100 or 200 years?

Scientists make large-scale climate predictions based on various three-dimensional mathematical computer models. Any computer model has to begin with certain assumptions, such as population, economic growth, and energy use.[41] Some predictions are based on a continuation of current trends, assuming that human activities do not change. Current IPCC predictions state that if no changes are made in human behavior, Earth's average global temperature will rise between 2.5°F and 10.4°F this century.[42]

Scientists know that greenhouse gases persist in the atmosphere for many years, which allows predictions to be based on current atmospheric levels. Yet, scientists do not fully understand the way ecosystems will respond to environmental changes. For example, the way carbon will be cycled in an ocean affected by pollution or increased global temperatures remains unclear. Given these unknowns, climate predications are general approximations. What most scientists agree upon is that global climate change is happening.

HISTORY OF THE ISSUE

Intergovernmental Panel on Climate Change

In 1988, the World Meteorological Organization (WMO) and the United Nations Environmental Programme (UNEP) established a panel to address the question of global climate change. The Intergovernmental Panel on Climate Change (IPCC) has grown to be regarded as "the international authority for conducting assessments of the current state of knowledge about climate change."[43] The IPCC makes climate-change predictions on the basis of assessments of peer-reviewed scientific literature.

The IPCC Second Annual Assessment provided scientific information used in crafting the 1997 Kyoto Protocol. The Third Assessment Report (available at the IPCC Web site) was issued in 2001, reflecting the opinions of more than 1,000 international experts. The fourth report will be issued in 2007.

United Nations Conference on Environment and Development (1992) "The Earth Summit"

In 1992, the United Nations hosted a summit in Rio de Janeiro, known as the Earth Summit, to develop international cooperation on environmental and economic concerns. Global climate change was a key agenda item, resulting in the formation of The Framework Convention on Climate Change, a legally binding agreement signed by 154 governments. Its ultimate objective is the "stabilization of greenhouse gas concentrations in the atmosphere at a level that would prevent dangerous anthropogenic (man-made) interference with the climate system."[44]

Kyoto Protocol

Five years after the 1992 Framework Convention on Climate Change was signed, a convention to set more rigorous emission-reduction standards was held in Kyoto, Japan. In 1997, countries set new goals to stabilize greenhouse concentrations "at a level

that would prevent dangerous anthropogenic interference with the climate system."[45]

Montreal Protocol

In 1987, an international agreement was adopted in which industrial nations that chose to sign the agreement were required to reduce their use of ozone-depleting chemicals, principally chlorofluorocarbons (CFCs). More than 180 countries have ratified the agreement, and many of the chemicals have been successfully phased out. The goal is to eliminate the use of CFCs, also a greenhouse gas, by 2010.[46]

Clean Air Act

In 1970, the United States passed the Clean Air Act (CAA), identifying and setting standards for air pollutants. The U.S. Environmental Protection Agency (EPA) was given the task of implementing the CAA in a way that would protect the public health and the environment. In 1977, the act was amended to include the New Source Review (NSR) permitting program. The amendment was to include older power plants in the regulations when they expanded or made significant modifications to their facilities.

CURRENT ISSUES AND FUTURE CONSIDERATIONS

Although some argue that global climate change is not a concern, many feel that we should heed the alarm call of thousands of international scientists. A majority of scientists believes that the rise in greenhouse gases is caused by human activity and that further change is inevitable.[47]

What Are the Economic and Environmental Concerns?

Continued warming trends in global temperatures will deepen already observed patterns of change. Rising sea levels, changes in major ocean currents, shifting patterns of rainfall and drought, health effects, and changes in ecosystems all have economic costs. Rising sea levels caused by the melting of glaciers and ice caps will flood coastal areas. On the basis of current levels of greenhouse gases in the atmosphere, the IPCC projects that the sea level will rise 1.5 to 3 feet (.46–.91 m) over the next 100 years.[48] About half of the population in the United States lives within 50 miles (80.4 km) of a coast. More than a billion dollars' worth of infrastructure—roads, buildings, and power lines—is located in coastal areas.[49] As an example of both the economic and environmental costs, the predicted rise in sea level could inundate the Everglades, thus flooding a world heritage site, an international biosphere reserve, and a wetland of international importance. The sea level rise would also make much of Florida uninhabitable.

In addition, melting of the polar ice caps and glaciers will increase global warming, both regionally and globally, by reducing the albedo, or heat reflected from the ocean surface.[50] Melting glaciers also will trigger shifts in weather patterns as the fresh, warmer waters disrupt the ocean currents, including the Gulf Stream. These weather changes could force hundreds of millions of people to move from their homes.[51]

Some scientists believe that global climate change is causing hurricanes to intensify in strength. In addition, changes in weather will affect agricultural lands and fishing stocks, along with the economic changes from floods, droughts, and disruption in ocean currents.

Loss of species and their habitats is another concern. The IPCC states in the 2002 report *Climate Change and Biodiversity*:

> Increasing global mean surface temperature is very likely to lead to changes in precipitation and atmospheric moisture because of

> changes in atmospheric circulation, a more active hydrological cycle ...[52]

Coral reefs and fisheries are already being damaged by warming waters. A study published in the scientific journal *Nature* in 2004 stated that if global-warming emissions are not reduced, more than a million species could be lost to extinction by 2050.[53]

Another concern is the increased threats to human health. The World Health Organization currently estimates that at least 160,000 human deaths annually are the result of climate change.[54] The EPA has warned that human health could be affected by increased heat and air pollution. Especially vulnerable are the young, the old, and those with heart and respiratory concerns. In addition, warmer weather could cause the spread of infectious diseases, such as malaria.[55]

No single approach will solve the problems of global climate change. Scientific understanding of the major causes, such as the increase in greenhouse gases in the atmosphere, is a strong beginning. Science has shown that burning fossil fuels adds significantly to the global carbon-dioxide levels, as does deforestation. According to Conservation International, one-fourth of our target goal for carbon reductions could be met if we stop cutting down forests.[56]

What Are the Policy Issues?

Much scientific opinion tells us that the potential effects of global climate change need to be addressed. The IPCC has stated the following:

> ... policies that lessen pressures on resources, improve management of environmental risks, and increase the welfare of the poorest members of society can simultaneously advance sustainable development and equity, enhance adaptive capacity, and reduce vulnerability to climate and other stresses.[57]

Such plans take two forms: mitigation and adaptation. Mitigation includes emissions reduction or reforestation plans, which is being addressed at many levels by some government and nongovernmental groups. Adaptation plans are developed to protect people and habitats from the results of climate change. As an example, Boston, Toronto, and London are involved in such planning, as well as agencies in New York and New Jersey.[58] The World Wildlife Fund and The Nature Conservancy are also involved in international projects that address adaptation strategies.[59]

Kyoto Protocol

On February 16, 2005, the Kyoto Protocol came into force, seven years after being negotiated as an extension of the 1992 UN Framework Convention on Climate Change. The Kyoto Protocol is a legally binding international agreement to place limits on greenhouse-gas emissions. For the 140 countries that signed and all others concerned about reducing the global atmospheric carbon, it was a reason for hope. The agreement states that the ratifying countries agree to reduce their combined emissions of six major greenhouse gases during the five-year period from 2008 to 2012 to below 1990 levels.[60]

Yet, the nation that emits about 25% of the total world carbon load but has only 5% of the world's population still refuses to sign the agreement to reduce emissions.[61] In 2001, United States president George W. Bush refused to endorse the agreement, and he again refused in 2005, maintaining that cutting carbon emissions would inhibit economic development. Many disagree, including 2,500 distinguished economists, eight of whom are Nobel laureates, who signed a statement in 1997 affirming that carbon dioxide and other greenhouse gases could be "cut in ways that would not reduce American living standards, could increase productivity, and would produce benefits that clearly outweighed the costs."[62]

Clean Air Act

A 1977 provision called the New Source Review was added to the Clean Air Act mandating that all stationary sources of air pollution be required to have permits before they are constructed or expanded. Some older power plants have used loopholes in the regulation to avoid making changes, which has resulted in lawsuits filed by states, cities, the American Lung Association, and environmental groups. Power plants remain an area of concern because they are the single largest generator of greenhouse gases. They are responsible for 37% of the carbon-dioxide emission from burning fossil fuels.[63]

McCain-Lieberman Climate Stewardship Act

In 2003, The McCain-Lieberman Climate Stewardship Act (CSA), a bipartisan approach to addressing the problems of global warming, was introduced in the Senate. Although the bill was defeated, the close 43 to 55 vote signaled growing national support for efforts to curb greenhouse-gas emissions. The bill was reintroduced in 2005, and again, defeated. However, this time the Senate did pass a resolution putting on record a call for a mandatory limit on U.S. global-warming pollution.[64]

State, Local, and Business Initiatives

The problems of climate change are felt on many levels—international, national, and local. Because of significant economic and environmental impacts associated with global climate change, businesses and governments at all levels are taking action.

As Seattle mayor Greg Nickels said at the June 2005 U.S. Conference of Mayors, "We're not going to wait for the federal government to do something to prevent the production of greenhouse gases. We're going to step up and provide the leadership at the local level, city by city."[65] Mayor Nickels's Climate Protection Agreement received unanimous support from the meeting, and more than 160 mayors from across the country have signed the agreement.

In 2004, officials from eight states and New York City filed a lawsuit to force five utilities to reduce their carbon-dioxide emissions. New York City Corporation Counsel, Michael Cardozo, stated the reason for the suit:

> The City of New York has joined this action out of concern for the impacts that global warming will have on the City and its residents and as part of the Bloomberg Administration's commitment to maintaining a clean and sustainable New York.[66]

States are also implementing their own emissions standards. In 2003, Maine signed a law requiring the state to develop a climate-change action plan to reduce carbon-dioxide emissions to 1990 levels by 2010, to 10% below 1990 levels by 2020, and by as much as 75 to 80% below 1990 levels over the long term.[67,68]

Other states are following a similar path. In 2003, nine Northeast and Mid-Atlantic states endorsed an action plan called the Regional Greenhouse Gas Initiative (RGGI) to develop a program to reduce carbon-dioxide emissions from power plants in participating states, while maintaining affordable and reliable energy.[69]

Some businesses are taking responsibility for decreasing their carbon-dioxide emissions and improving their energy efficiency. One example is the worldwide business leaders who are members of Pew's Center Business Environmental Leadership Council. Members include Toyota, Hewlett-Packard, Boeing, DuPont, IBM, Weyerhaeuser, and Whirlpool. These companies are taking proactive and innovative approaches to setting targets for reducing greenhouse gases, participating in emissions trading, and improving energy efficiency.[70] In fact, Dupont has cut its greenhouse-gas emission by 72% from 1990 levels, while saving $2 billion in reduced energy bills.[71]

Another example of business initiative is the May 2005 announcement by General Electric (GE), the world's second-largest company, to cut its global-warming emissions and make

significant investments in clean energy technologies. GE also joined other companies in asking Congress to set national targets to reduce global-warming emissions.

Nongovernmental Organizations

Environmental organizations across the globe are working to forge partnerships with other nongovernmental organizations, businesses, and governments to reduce global warming. In 2002, Conservation International (CI) began its Conservation Carbon Initiative to develop "offset projects," or land-based projects to reduce carbon-dioxide levels by stopping deforestation or by replanting degraded lands. CI now has these projects in five countries with "hot spot" areas of biodiversity.[72] The World Wildlife Fund and The Nature Conservancy also develop partnerships with businesses and governments to protect forests. In 2004, the World Wildlife Fund developed the Climate Savers Program to help businesses innovate climate-change and energy solutions, while gaining international recognition. The first seven companies to sign on include Johnson & Johnson, IBM, Polaroid, Nike, Lafarge, The Collins Companies (wood products), and Sagawa Express.[73]

WHAT THE INDIVIDUAL CAN DO

The problems of climate change can seem overwhelming to the individual. However, consider that about 30% of the production of greenhouse gases comes from sources that we, as individuals, can control. Americans use a lot of electricity, gas, and oil. According to one study, the energy used to operate the average American home adds more than 24,000 pounds (10,886.2 kilograms) a year of carbon dioxide to the atmosphere.

Here are some steps that can be taken by the individual to cut the emissions of greenhouse gases:

1. Make your home more energy efficient. Use energy-efficient appliances. Look for the "energy star" label on new appliances. If each household used more efficient appliances, we could save an estimated $15 billion in energy and reduce greenhouse-gas emissions by 175 million tons (158 million metric tons) a year.[74]
2. Drive more fuel-efficient cars. A car that gets 20 miles (32.2 km) per gallon emits twice as much carbon dioxide as one that gets 40 miles (64.4 km) per gallon.[75]
3. Buy recycled and recycle. Recycling can reduce your home's carbon-dioxide emissions by 850 pounds (385.5 kg) per year. Buying food in recyclable containers or reduced packaging can cut emissions by 230 pounds (104.3 kg) a year.[76] For example, 95% of the energy used to produce an aluminum can is conserved when the can is recycled.[77]

REFERENCES

1. "Global climate change initiative," The Nature Conservancy. Available online. URL: nature.org/initiatives/climatechange. 2005.
2. Hardy, John T. *Climate Change: Causes, Effects, and Solutions.* Hoboken, N.J.: John Wiley and Sons, 2003.
3. "The heat is on: A white paper on climate change," Environmental Defense. Available online. URL: http://www.environmentaldefense.org/documents/3777_TheHeatIsOn.pdf. 2004.
4. Hardy, *Climate Change*, p. 11.
5. Third Annual Report. "Climate change 2001," Intergovernmental Panel on Climate Change. Available online. URL: http://www.grida.no/climate/ipcc_tar/wg1/index.htm.
6. Turco, Richard P. *Earth Under Siege: From Air Pollution to Global Change*. New York: Oxford University Press, 2002.
7. Starr, Cecie, and Ralph Taggart. *Biology: The Unity and Diversity of Life*. 8th Edition. Belmont, Calif.: Wadworth Publishing, 1998.
8. Hardy, *Climate Change*, p. 10.
9. Blatt, Harvey. *America's Environmental Report Card: Are We Making the Grade?* Cambridge, Mass.: MIT Press, 2005.
10. Factsheet. "Global warming: Emissions," Environmental Protection Agency. Available online. URL: http://yosemite.epa.gov/OAR/globalwarming.nsf/content/Emissions.html. 2005.
11. "Global environment," Union of Concerned Scientists. Available online. URL: http://www.ucsusa.org/global_environment/global_warming/page.cfm?pageID=497. 2005.
12. IPCC Third Annual Report. "Climate change 2001: A scientific basis," Available online. URL: http://www.grida.no/climate/ipcc_tar/wg1/028.htm#e8. 2001.
13. Conniff, Richard. "Counting carbons: How much greenhouse gas does your family produce?" *Discover* 26 (2005): 56.
14. Factsheet. "Global warming: Emissions," Environmental Protection Agency. Available online. URL: http://yosemite.epa.gov/oar/globalwarming.nsf/content/emissionsindividual.html. 2000.
15. Starr, *Biology*, p. 889.

16. Sawin, Janet. *Climate Change Indicators on the Rise.* In *Vital Signs.* New York: Norton, 2005.
17. "Climate change," World Wildlife Fund. Available online. URL: www.worldwildlife.org/climate/basic.cfm. 2005.
18. New York Climate Change Information Resources. "What are the causes of global warming?" Available online. URL: http://ccir.ciesin.columbia.edu/nyc/ccir-ny_q1b.html. 2005.
19. Wilson, E.O. *The Future of Life.* New York: Vintage Books, 2002.
20. "Climate change science: An analysis of some key questions," National Research Council. Washington DC: National Academy Press. Available online. URL: http://yosemite.epa.gov/OAR/globalwarming.nsf/UniqueKeyLookup/SHSU5BUTQ4/$File/nas_ccsci_01.pdf. 2001.
21. Factsheet. "Global warming: Climate," Environmental Protection Agency. Available online. URL: http:// Yosemite.epa.gov/OAR/globalwarming.nsf/content/Climate.html. 2005.
22. Sawin, *Vital Signs*,. p. 40.
23. "Global warming and our changing climate," US Environmental Protection Agency. 430-F-00-011, 2000.
24. Pearce, R. "And now the weather: Rain, rain, here to stay." *New Scientist* (November 1997): 28.
25. "Confronting climate change in the Great Lakes region," Union of Concerned Scientists and the Ecological Society of America. Available online. URL: http://www.ucsusa.org/greatlakes/glchallengetoc.html. 2002.
26. Hassol, Susan Joy. *Arctic Climate Impact Assessment.* Cambridge, UK: Cambridge University Press, 2004.
27. IPCC Third Annual Report. "Climate change 2001: A scientific basis," Available online. URL: http://www.grida.no/climate/ipcc_tar/wg1/028.htm#e8. 2001.
28. Fishetti, M. "Drowning New Orleans." *Scientific American*, October 2001: 78.
29. "The tide is higher: Study shows rise of Atlantic sea level." *Bulletin of the American Meterological Society*, February 2002: 170.
30. "Global warming: Glacier National Park is a global warming laboratory," Sierra Club Factsheet. Available online. URL: http://www.sierraclub.org/globalwarming/articles/glacier.asp.

31. "Glaciers on Mount Rainier," National Park Service Available online. URL: http://www.nps.gov/mora/ncrd/glaciers.htm. 2005.
32. "Climate change science: An analysis of some key questions," National Research Council. Washington DC: National Academy Press. Available online. URL: http://yosemite.epa.gov/OAR/globalwarming.nsf/UniqueKeyLookup/SHSU5BUTQ4/$File/nas_ccsci_01.pdf. 2001.
33. "Alarming meltdown at Ilulissat Icefjord world heritage site," IUCN WCPA New and Events. Available online. URL: http://www.iucn.org/themes/wcpa/newsbulletins/new.htm. March 2005.
34. "Climate clues from glaciers," *Science* 308 (2005): 597.
35. Spray, Sharon and Karen McGlothlin (eds). *Loss of Biodiver*sity. New York: Rowman & Littlefield, 2002.
36. Semlitsch, Raymond D., ed. *Amphibian Conservation*. Washington D.C.: Smithsonian Books, 2003.
37. Nickens, T. Edward. "North American fish feel the heat," National Wildlife Federation. Available online. URL: http://www.nwf.org/nationalwildlife/dspPlainText.cfm?articleId=503. June/July 2002.
38. Agary, Tundi. "America's coral reefs: Awash with problems." *Issues in Science and Technology*. Available online. URL: http://www.issues.org/20.2/agardy.html. 2003.
39. "Climate change: The threat," The Nature Conservancy. Available online. URL: http://nature.org/initiatives/climatechange/about/. 2005.
40. "The problem of land degradation," United Nations Convention to Combat Desertification. Available online. URL: http://www.unccd.int/. 2005.
41. "Global warming" National Oceanic and Atmospheric Administration. Available online. URL: http://www.lwf.ncdc.noaa.gov/oa/climate/globalwarming.html. 2005.
42. Wilson, E.O. *The Future of Life*. New York: Vintage Books, 2002.
43. IPCC Press Advisory. Geneva/Paris. February 14, 2004.
44. United Nations Department of Economic and Social Affairs: Division for Sustainable Development. UNCED Factsheet. Available online. URL: http://www.un.org/esa/sustdev/sdissues/water/Interagency_activities.htm. 2005.

45. Johannesburg Summit 2002. UNCED. Available online. URL: http://www.johannesburgsummit.org/html/basic_info/unced.html. 2002.
46. "History of Kyoto Protocol," Pew Center for Climate Change. Available online. URL: http://www.pewclimate.org/history_of_kyoto.cfm. 2005.
47. "The Vienna convention and the Montreal protocol," United Nations Development Program. Available online. URL: http://www.undp.org/seed/eap/montreal/montreal.htm. 2005.
48. "The science of global warming," Union of Concerned Scientists. Available online. URL: http://www.ucsusa.org/global_warming/science/. 2005.
49. "Global warming and our changing climate," Environmental Protection Agency. 430-F-00-011, 2000.
50. "Climate change science: An analysis of some key questions," National Research Council. Washington D.C.: National Academy Press. Available online. URL: http://yosemite.epa.gov/OAR/globalwarming.nsf/UniqueKeyLookup/SHSU5BUTQ4/$File/nas_ccsci_01.pdf. 2001.
51. Hassol, *Arctic Climate Impact Assessment*, p. 30.
52. "The heat is on: A white paper on climate change," Environmental Defense. Available online. URL: http://www.environmentaldefense.org/documents/3777_TheHeatIsOn.pdf. 2004.
53. "Climate change and biodiversity," IPCC Technical Paper V. Intergovernmental Panel on Climate Change. Available online. URL: http://www.ipcc.ch/pub/tpbiodiv.pdf.
54. Sawin, *Vital Signs*.
55. "Global warming and our changing climate," Environmental Protection Agency. 430-F-00-011, 2000.
56. "Climate change," Conservation International. Available online. URL: http://www.conservation.org/xp/CIWEB/programs/climatechange. 2005.
57. Intergovermental Panel of Climate Change. "Climate 2001: Impacts, adaptation, and vulnerabilty," Available online. URL: http://www.grida.no/climate/ipcc_tar/. 2001.
58. Climate Change Information Resources. New York Metropolitan Region. "Adaptation: Public sector strategies," Available online.

URL: http://www.ccir.ciesin.columbia.edu/nyc/ccir-ny_q3.html. 2005.

59. "Our solution: Adaptations," World Wildlife Fund. Available online. URL: http://www.panda.org/about_wwf/what_we_do/climate_change/our_solutions/adaptations.cfm. 2005.
60. UN News Center. "UN's Kyoto Treaty against global warming comes into force," Available online. URL: http://www.un.org/apps/news/story.asp?NewsID=13359&Cr=global&Cr1=warm#. February 16, 2005.
61. Sawin, *Vital Signs*, p. 40.
62. Shabecoff, Philip. *Earth Rising*. Washington, DC: Island Press, 2003.
63. "Power companies fail to chart clear course to combat climate change," Newsroom: World Wildlife Press Release. Available online. URL: http://www.worldwildlife.org/news/displayPR.cfm?prID=166. November 30, 2004.
64. "Senate says: U.S. must enact mandatory limits on global warming pollution," Natural Resources Defense Council Press Release. Available online. URL: http://www.nrdc.org/media/default.asp#0624.
65. "U.S. Mayors endorse Nickels' climate protection agreement," News Advisory. City of Seattle. Available online. URL: http://www.cityofseattle.Net/news/detail/asp?ID=5260&Dept=40. June 13, 2005.
66. "Eight states and NYC sue top five U.S. global warming polluters," Press Releases. Office of New York State Attorney General Elliot Spitzer. Available online. URL: http://www.oag.state.ny.us/press/2004/jun/jun28b_04.html. July 2004.
67. State News Archives. Pew Center for Climate Change. Available online. URL: http://www.pewclimate.org/what_s_being_done/in_the_states/2004archives.cfm. 2004.
68. "New Mexico Governor Richardson joins global warming action outside," NRDC Press Archive: Washington. Available online. URL: http://www.nrdc.org/media/pressreleases/050609b.asp. June 19, 2005.
69. "Regional greenhouse gas initiative," Factsheet. Available online. URL: www.rggi.org/about.htm. 2005.

70. Business Environmental Leadership Council. Pew Center for Global Climate Change. Available online. URL: http://www.pewclimate.org/companies_leading_the_way_belc/.
71. Conniff, Richard. "Counting carbons: How much greenhouse gas does your family produce?" *Discove*r 26 (2005): 56.
72. Frontlines Online. "Engaging business in CO_2 clean-up" Available online. URL: http://www.conservation.org/xp/frontlines/partners/08110414.xml. 2005.
73. "Climate change: Featured projects," World Wildlife Fund. Available online. URL: http://www.worldwildlife.org/climate/projects/climateSavers.cfm. 2005.
74. "Global warming: Emissions," Environmental Protection Agency. Available online. URL: http://www.yosemite.epa.gov/oar/globalwarming.nsf/content/emissionsindividual.html.
75. "Climate solutions," Union of Concerned Scientists. Available online. URL: http://www.ucsusa.org/global_environment/global_warming/page.cfm?pageID=1413. 2005.
76. "How to fight global warming," Natural Resources Defense Council. Available online. URL: http://www.nrdc.org/globalWarming/gsteps.asp. 2003.
77. "Make a difference at the store," Environmental Protection Agency. Available online. URL: http://www.yosemite.epa.gov/OAR/globalwarming.nsf/content/ActionsIndividualMakeaDifference.html. 2000.

FURTHER READING

Books

Calhoun, Yael. Ed. *Environmental Issues: Climate Change.* New York: Chelsea House Publishers, 2006.

Gore, Al. *An Inconvenient Truth.* Emmaus, Penn.: Rodale Books, 2006.

Hardy, John T. *Climate Change: Causes, Effects, and Solutions.* Hoboken, N.J.: John Wiley and Sons, 2003.

Web Sites

The Nature Conservancy Climate Change Initiative

http://www.nature.org/initiatives/climatechange/

The World Wildlife Fund

http://www.worldwildlife.org/climate/index.cfm

GLOSSARY

Aerosols Microscopic solid particles suspended in the atmosphere, such as dust, soot, or pollutants.

Bioaccumulate The process by which a substance, usually a toxic chemical, collects in the tissue of a living organism over time. An animal that eats the organism in which the chemicals have accumulated also absorbs and collects the toxins in its body.

Biodiversity The diversity of living things; the term refers to diversity on three levels: ecosystem diversity, species diversity, and genetic diversity.

Carbon sinks Mechanisms that absorb carbon, such as forests and oceans.

Carcinogen A substance that causes cancer.

Combustion A sequence of chemical reactions that results in the production of heat and light. The process results in certain by-products, including unburned carbon and carbon compounds.

Condensation The changing of a gas to a liquid.

Decompose To break down into basic components, such as when dead animals or plants break down into minerals and nutrients.

Ecosystem The interaction among the living (plants, animals, and microorganisms) and the nonliving (soil, air, weather, water, and sunlight) components in an area.

Endangered species Living organisms in danger of becoming extinct.

Erosion The process by which the forces of water, wind, ice, rocks, or soils wear away or transport rocks and soil.

Evaporation The change in the state of a substance from liquid to gas.

Food chain A linear representation of the way in which organisms feed upon one another.

Fossil fuels Fuel such as coal, petroleum, and natural gas that is formed from the remains of ancient plants and animals.

Fuel cell A device that generates electricity by combining hydrogen and oxygen into water.

Greenhouse gas Gas in the atmosphere that traps heat energy and contributes to the warming of Earth's atmosphere.

Hydrocarbons Compounds composed of hydrogen and carbon, most commonly found in fossil fuels. The most basic example of a hydrocarbon is methane.

Moratorium A legal action declaring a temporary delay of a particular activity.

Oxidation The chemical process of combining oxygen with another substance in which electrons are lost.

Ozone A molecule that consists of three atoms of oxygen. It is considered a pollutant in the lower atmosphere but is essential in the upper atmosphere because it creates a protective layer around Earth.

Particulate matter (PM) A pollutant in the atmosphere that can take the form of both a solid particle and a liquid droplet. Particulate matter is 100 micrometers in diameter (0.0039 inches) or less.

Photosynthesis The chemical process by which light energy from the sun powers a reaction that chemically changes water and carbon dioxide into glucose.

Pollination The first step in seed production; it is the transfer of pollen from the stamen to the stigma via wind, water, animals, or humans.

Terrarium An enclosure, usually a clear container, containing some combination of water, soil, plants, and animals for the purpose of observing living organisms in their natural environmental conditions.

Wetland Land that has standing water for certain periods during the year, hydric (wetland) soils, and supports wetland vegetation.

PICTURE CREDITS

1: U.S. Fish and Wildlife Service
3: Associated Press
7: Infobase Publishing
9: Infobase Publishing
11: Terry Donnelly/Dembinsky Photo Associates
16: U.S. Fish & Wildlife Service
19: U.S. Fish & Wildlife Service
26: Sharon Cummings/Dembinsky Photo Associates
35: AbleStock.com
37: Scott T. Smith/Dembinsky Photo Associates
41: Associated Press
45: Associated Press
47: Associated Press
48: Corbis
54: Associated Press
65: Willard Clay/Dembinsky Photo Associates
68: Infobase Publishing
69: Infobase Publishing
70: Infobase Publishing
72: Infobase Publishing
74: Infobase Publishing
75: Infobase Publishing
78: Getty

Cover: AbleStock.com

INDEX

INDEX

INDEX

= ABOUT THE AUTHOR =

YAEL CALHOUN is a graduate of Brown University. She has an M.A. in education and an M.S. in natural resources science. Years of work as an environmental planner have provided her with much experience in environmental issues at the local, state, and federal level. Currently, she writes books and teaches environmental biology at Westminster College in Salt Lake City. She lives with her family at the foot of the Rocky Mountains in Utah.